AF725589

Technocracy and Democracy in Latin America

Praised by some as islands of efficiency in a sea of unprofessional, politicized, and corrupt states, and criticized by others for removing wide-ranging areas of policy making from the democratic arena, technocrats have become prominent and controversial actors in Latin American politics. Nonelected state officials with advanced educations from top universities, technocrats achieved considerable autonomy from political and economic actors and exerted great influence over their countries' fates. This finding poses an intriguing paradox. These experts lack an independent base of authority, such as that afforded via popular election. They also lack the tenure enjoyed by professional bureaucrats. What, then, explains the power of technocrats in democratic Latin America? And why do they enjoy and maintain greater influence in some policy areas than in others? Through an in-depth analysis of economic and health policy in Colombia from 1958 to 2013 and in Peru from 1980 to 2013, *Technocracy and Democracy in Latin America* answers these and other questions about technocrats in Latin America.

Eduardo Dargent is an associate professor of political science at Pontificia Universidad Católica del Perú. His main teaching and research interests are comparative public policy and democratization, and the state in the developing world. He has published in *Comparative Politics*, the *Journal of Latin American Studies*, and the *Journal of Politics in Latin America*.

Technocracy and Democracy in Latin America

The Experts Running Government

EDUARDO DARGENT
Pontificia Universidad Católica del Perú

CAMBRIDGE
UNIVERSITY PRESS

CAMBRIDGE
UNIVERSITY PRESS

Shaftesbury Road, Cambridge CB2 8EA, United Kingdom

One Liberty Plaza, 20th Floor, New York, NY 10006, USA

477 Williamstown Road, Port Melbourne, VIC 3207, Australia

314–321, 3rd Floor, Plot 3, Splendor Forum, Jasola District Centre, New Delhi – 110025, India

103 Penang Road, #05–06/07, Visioncrest Commercial, Singapore 238467

Cambridge University Press is part of Cambridge University Press & Assessment, a department of the University of Cambridge.

We share the University's mission to contribute to society through the pursuit of education, learning and research at the highest international levels of excellence.

www.cambridge.org
Information on this title: www.cambridge.org/9781107059870

© Eduardo Dargent 2015

This publication is in copyright. Subject to statutory exception and to the provisions of relevant collective licensing agreements, no reproduction of any part may take place without the written permission of Cambridge University Press & Assessment.

First published 2015

A catalogue record for this publication is available from the British Library

Library of Congress Cataloging-in-Publication data
Dargent, Eduardo, 1974–
Technocracy and democracy in Latin America : the experts running government / Eduardo Dargent.
pages cm
Includes bibliographical references and index.
ISBN 978-1-107-05987-0 (hardback)
1. Latin America – Politics and government – 1948–1980. 2. Latin America – Politics and government – 1980– 3. Technocracy. 4. Government executives – Latin America. 5. Democracy – Latin America. 6. Latin America – Economic policy. 7. Medical policy – Latin America. I. Title.
JL960. D38 2015
320.98–dc23 2014033135

ISBN 978-1-107-05987-0 Hardback

Cambridge University Press & Assessment has no responsibility for the persistence or accuracy of URLs for external or third-party internet websites referred to in this publication and does not guarantee that any content on such websites is, or will remain, accurate or appropriate.

Contents

Tables

Acknowledgments

Some parts of this work (most notably sections of Chapters 2, 4, and 5) were published in "Agents or Actors? Assessing the Autonomy of Economic Experts in Colombia and Peru," *Comparative Politics*, Vol. 43 No. 3 (April 2011).

I have incurred many debts throughout the writing process. This book is based on my doctoral thesis so, as you can imagine, there is a long list of individuals to whom I owe words of gratitude. First and foremost, I am grateful for the generous support of my academic adviser, Kurt Weyland. Throughout the years I spent working on my thesis, Kurt promptly returned everything I presented to him with detailed and thoughtful comments. I truly appreciate his generosity as an adviser, and I hope to be able to help my students in a similar capacity. I am also grateful to other members of my dissertation committee: Dan Brinks, Raúl Madrid, and Wendy Hunter. They provided me with invaluable insight and excellent feedback, even when I was no longer under their tutorship. In general, the fantastic clout of Latin American specialists, professors, and students at the Department of Government at The University of Texas at Austin helped craft the perfect environment for me to carry out this work.

There are many other individuals in the Department of Government at UT Austin who deserve acknowledgment for their inestimable support, including Manuel Balan, Catherine Boone, Sandra Botero, Daniel Budny, Luis Camacho, Annette Carlile, Henry Dietz, Laura Field, Austin Hart, Ilana Lifshitz, Robert Moser, Rodrigo Nunes, Daniel Ryan, and Mary Slosar. All were generous friends who provided support both inside and outside the classroom. More generally, my research would not have been possible without a number of grants and scholarships provided by institutions at UT Austin: the Department of Government, the Teresa Lozano Long Institute of Latin American Studies, and the College of Liberal Arts.

I also would like to express my gratitude to the many individuals who helped me during my research. Diverse academics, public officers, NGO activists, journalists, and, of course, technocrats, were generous with their valuable time. There are too many to mention, so I will only briefly list some here. In Colombia, Felipe Botero, Ralf Leiteritz, Rodolfo Masías, Carlo Nasi, and Laura Wills from the Political Science Department of Universidad de Los Andes provided me with numerous contacts and invaluable advice. Angelika Rettberg, the director of the department, kindly organized two workshops centered on my work, an internal one with professors and a second one open to the academic community. Both events helped me think through my research. In FEDESARROLLO, Mauricio Cárdenas, Mauricio Santa María, and Alejandra Gonzalez shared their knowledge about economics and public policy and provided me with valuable contacts. I also would like to give special thanks to Juan Pablo Uribe, health expert of the Fundación Santa Fé de Bogotá, who was patient enough to explain the Colombian health system, by far one of the most daunting challenges in my work. Diana Rodríguez Franco was also a great supporter, helping me fill in the holes in my previous research throughout this past year. In Colombia, my utmost gratitude goes to the two individuals largely responsible for my success in disentangling the political and technocratic culture of this fascinating country: Otto Morales Benites and Rodrigo Botero, a politician and a technocrat, respectively, who went out of their way to support me throughout this project, providing contacts, advice, and information. My admiration for these two statesmen is enormous.

In Peru, in addition to those I interviewed, I would like to recognize my friends and colleagues and the administrative staff at the Pontificia Universidad Católica del Perú (PUCP). Rolando Ames, Aldo Panfichi, and Catalina Romero made me feel like a part of this institution from day one. Jorge Aragón, José Carlos Orihuela, and Maritza Paredes provided inestimable information and comments on different parts of my work. Among my technocratic friends, María Inés Vásquez was a superb translator of "who is who" in the Peruvian state. Not only did she provide me with contacts that would have otherwise taken me months to find, but she also spent time learning about my interests and suggesting promising avenues of research. Richard Webb and Milagros Mejía from the Instituto del Perú of the Universidad San Martín gave me an office during my first period of field research in Lima. They also organized a workshop where I discussed my preliminary findings in the country. The Consorcio de Investigación Económica y Social (CIES) awarded me a research grant that helped me conduct research in the Ministries of Economy and Health.

Whether professors or PhD students, many other comparative scholars also deepened my research and writing experience. Both Ezequiel Gonzales and Jennifer Cyr conducted their field research at the same time as I did, which allowed for much conversation and debate. Steve Levitsky spent a year in PUCP while I was teaching there, and his energy and engagement became invaluable assets in our department. Juan Pablo Luna from Pontificia Universidad Católica de Chile and his colleagues were kind enough to organize a book workshop to discuss this manuscript in June 2013. My friend Hillel Soifer is a great academic who provides fabulous advice, and some of the names and concepts in this work are a direct result of his creativity and generosity. Alberto Vergara deserves a special word of recognition for the enormous support and positive influence he provided in both my personal and academic life. It has been more than two decades of friendship and collaboration since we started our two-person reading group about political science and theory. I would also like to thank Mary Sutcliffe Love, Aaron Watanabe, and Annette Wren, all of whom have helped me edit my Spanglish at different points throughout the process. Of course, all errors within this book are my responsibility.

Last but not least, my family provided much needed emotional support. Both my mother, Leticia Bocanegra, and father, Eduardo Dargent Chamot, have encouraged me to follow my academic interests since early childhood and have always supported my decisions. My brother Felipe, the real scientist in the family, and Gabriela Capurro, my bright and charming sister-in-law, are also important members of this tight support group. My generous and caring Colombian and Peruvian families gave me more encouragement than they could imagine. Lastly, my wife Paula Muñoz deserves to be mentioned in each of these paragraphs. A colleague, a friend, a travel companion, and above all a loving partner, there is more of Paula in this work than I can convey. Her love, care, intelligence, and sensibility have seen me through this trying adventure, one of many more to come.

1

Introduction

Technocracy under Democracy

In June 2005, Peruvian President Alejandro Toledo (2001–2006) complained bitterly about a "silent opposition" that prevented him from fulfilling his electoral promises. With his approval ratings in the low twenties and the 2006 general election approaching, Toledo had good reason to be worried. *Perú Posible*, his political party, was expected to lose all but two of its thirty-five congressional seats (30 percent of Congress) in the election, while its prospects for the regional and municipal elections scheduled for later that year were similarly bleak. Surprisingly, Toledo saw this "silent opposition" operating within the executive branch. According to the president, it consisted of a group of "insensitive bureaucrats" (*burócratas insensibles*) in the Ministry of Economy and Finance (MEF) that tied his hands and delayed public investment even in times of significant economic growth.[1] Some months before leaving office, Toledo told his economic team, "without technocrats I would have done much more." His two-time economic minister, and then cabinet chief, Pedro Pablo Kuczynski, answered playfully: "The 'insensitive technocrats' gave you 6.5 percent GDP growth per year, so stop whining."[2]

Why did Toledo publicly complain about his economic technocrats while giving them control over the national budget? If they were such an obstacle for him, why did he not just fire them? The president could have appointed members of his party to economic institutions and thereby have easily increased public investment in order to enhance *Perú Posible*'s electoral chances. After all, these *técnicos* were mostly mere appointees without bureaucratic tenure. Toledo's seemingly surprising decision to keep technocrats in their positions was not exceptional: for more than twenty years, economic technocrats have been in charge of economic policy in Peru. Despite their populist campaign styles,

[1] *La República*, "Toledo Critica a Funcionarios 'Intermedios' de su Gobierno," June 12, 2005.
[2] *Caretas*, "A Doce Mil Pies," February 9, 2006.

Toledo's successors, Alan García (2006–2011) and Ollanta Humala (2011–), kept similarly technically trained individuals in control of economic policy.

Nor was Toledo's decision an exception by regional standards. Beginning in the 1980s and intensifying during the neoliberal reform period of the 1990s, Latin American countries witnessed the appointment of economic technocrats to key state positions where, to this day, they promote and implement policies they prefer while frequently vetoing those they oppose. Praised by some as islands of efficiency in otherwise unprofessional, politicized, and corrupt states, and criticized by others for removing wide-ranging areas of policy making from the democratic arena, economic technocrats have become prominent and controversial actors in Latin American politics (P. Silva 1991; Dominguez 1997; Conaghan and Malloy 1994; Teichman 1997; Centeno and Silva 1998; Babb 2005; Montecinos 2005). Some commentators have even discussed the appearance in the region of what they call "technocratic democracies," defined as regimes where the "framing of policy alternatives is largely in the hands of experts" (Centeno and Silva 1998, 11).

Nonetheless, the power wielded by economic experts is not always matched by technocrats elsewhere in the state. Between 1993 and 2002, a group of health policy technocrats accumulated significant influence within the Peruvian Ministry of Health (Ministerio de Salud–MINSA). Within a ministry characterized by immobility and bureaucratic weakness, these experts enacted significant reforms, often without direct involvement from politicians. These technocrats were, however, finally dismissed when the externally funded reform projects they oversaw ran out of funds and Toledo's first health minister rejected their approach to public health. In clear contrast to the ensuing media spectacle every time the president appoints a new finance minister, these events went largely unnoticed by politicians and the press. A similar pattern of economic experts, enjoying more job stability than do technocrats in other areas such as health policy, is also observed in Colombia and other Latin American countries.

Why have economic experts in the MEF achieved considerable autonomy and influence for two decades in Peru while health experts failed to gain stability in MINSA? Where does experts' power over policy making come from if they lack an independent base of authority, such as that afforded via popular election, or the tenure enjoyed by professional bureaucrats? Why do technocratic reform teams enjoy and maintain greater autonomy in certain policy areas than in others? More broadly, are these non-elected state officials a positive or negative influence on democracy in the region? Through an in-depth analysis of technocratic experiences in economic and health policy in Colombia from 1958 to 2013 and in Peru from 1980 to 2013, this book aims to answer these and other questions about technocrats in Latin America.

THE PUZZLE

Technocrats are long-standing actors in Latin America. Appointing well-qualified individuals to state positions is a way to achieve greater capacity than the region's traditionally weak bureaucracies could provide (Evans 1992, 141; Geddes 1994). Historically, *técnicos* were associated with authoritarian and semi-authoritarian regimes. Under these regimes, experts were insulated from popular and electoral pressures, which allowed for relatively quick and unimpeded implementation of their preferred policies. For example, Guillermo O'Donnell's concept of bureaucratic authoritarianism highlighted the importance of high-ranking bureaucrats in the military governments of the Southern Cone during the 1970s (O'Donnell 1973; Collier 1979; Biglaiser 2002). Experts also implemented ambitious reforms during the military governments of the 1960s and 1970s in Brazil (Geddes 1990; Hagopian 1994). Further examples include the highly trained bureaucrats who challenged landed and industrial elites under Peru's military regime following the 1968 "revolution from above," (Stepan 1978; Trimberger 1978); the market-oriented "Chicago Boys" behind Augusto Pinochet's neoliberal transformation of Chile (O'Brien and Rodick 1983; P. Silva 1991); and, more recently, the experts linked to the hegemonic *Partido Revolucionario Institucional* (PRI) who carried out market reforms during Carlos Salinas's (1988–1994) and Ernesto Zedillo's (1994–2000) presidencies in Mexico (Centeno 1994; Teichman 1997; Babb 2005).

For most of the past century, however, technical expertise seemed scarce in Latin American democracies. Patronage-driven politicians in democratic Latin America were frequently unwilling to create areas of technical competence that would impose restraints on their use of resources (Malloy 1992; Webb 1994). However, since the 1960s, economic technocrats have become important actors in many Latin American democracies, especially during recent periods of neoliberal reform. In Colombia, technocrats have been key actors since the late 1950s when President Alberto Lleras Camargo (1958–1962), and more clearly, President Carlos Lleras Restrepo (1966–1970), strengthened the executive branch by appointing experts in several economic institutions. More recently, experts became top players in Peruvian politics during Alberto Fujimori's regime (1990–2000). *Técnicos* have also remained key policy makers in Chile and Brazil following the countries' democratic transitions from authoritarian rule in the late 1980s (P. Silva 2008). Similarly, the PRI's failure to maintain its electoral hegemony in 2000 did not reduce the importance of economic experts in the Mexican state (Babb 2005).

Experts' increased role under democratic regimes has stimulated a lively theoretical debate about their influence in policy making and effective independence from other sociopolitical players, especially politicians, international financial institutions (IFIs), and business interests. What explains the relevance of experts in contemporary Latin American democracies? Is their power derived from other actors or do they enjoy autonomy? Three instrumentalist theories locate the source of technocrats' influence in other political or economic actors. These instrumentalist theories conclude that *técnicos* are "merely representatives or the most visible manifestation of deeper or more structural determinacies; technocrats are not where the action is" (Centeno and Silva 1998, 2).

As discussed in Chapter 2, these instrumentalist theories fail to capture the importance of experts in the region. Even if these theories provide important insights into the role of technocrats, they ultimately underestimate the autonomy experts possess from their alleged principals and the capacity to advance their own policy goals. Following previous works, I propose that technocrats are better conceptualized as autonomous actors (Centeno and Silva 1998; P. Silva 2008). Experts fulfill the most influential definition of state autonomy in the comparative politics literature: "formulate and pursue goals that are not simply reflective of the demands or interests of social groups, classes, or society" (Skocpol 1985, 9).

In the following pages, I demonstrate how technocratic reform teams frequently adopt their own policy preferences, which they advance with relative success, even against the interests of other powerful actors, including their democratically elected superiors. In some policy areas, the implementation of experts' preferred policies significantly alters society, deeply transforming institutional arrangements and/or the distribution of resources among social groups.[3] Furthermore, in some areas *técnicos* even achieve entrenchment – meaning incumbent politicians are informally forced to appoint officials in that policy area with similar levels of training and policy preferences – that endures over time. Appointing unqualified individuals in directive positions for these policy areas carries high political costs for the incumbent. As such, these highly educated state officials meet the criteria for autonomy specified in previous studies of states, international organizations, and bureaucratic agencies (Krasner 1984; Brint 1990, 369; Centeno 1993, 322–323; Barnett and Finnemore 1999; Carpenter 2001, 15).

These findings pose an intriguing paradox. Experts lack an independent base of power, such as that afforded via popular election, or institutional protections as in the case of certain state bodies (courts, central banks, etc.). They also lack the tenure enjoyed by professional bureaucrats because usually they are merely appointees. Formally, most experts remain highly dependent on politicians for their positions, as they "can be fired in a matter of minutes" (Schneider 1998, 78). How, then, can technocrats be autonomous and influential in their policy areas? What is the source of their power? Moreover, why does this autonomy vary across policy sectors and even within the same policy sector over time?

Through an in-depth analysis of economic and health technocrats in Colombia and Peru, I propose a theory of technocratic autonomy that develops and refines our understanding of the power of experts in the region. This theory explains why experts can achieve significant autonomy, illuminates the determinants of variation in the degree of this autonomy across these policy areas and across time, and elucidates why technocratic power becomes entrenched in certain areas, resulting in experts essentially running government within their

[3] Brint uses these criteria to highlight the power of bureaucratic actors (Brint 1990, 369).

policy domains. This theory also provides insights into the conditions that can lead to the fall of such powerful state officials. In this introductory chapter, I briefly present this theory concerning technocrats in Latin America. Next, I explain the theoretical relevance of this analysis. I conclude by specifying two central concepts in this work – technocrats and democracy– and providing an overview of the book.

THE ARGUMENT IN BRIEF

Technocrats' main source of autonomy is expertise, which legitimizes *técnicos* and their preferred policies while also giving them considerable leverage over sociopolitical actors. Technocrats are usually appointed to deal with complex issues that *políticos* lack the knowledge to successfully address. My study offers evidence of how, once in place, technical reform teams often adopt decisions that diverge from politicians' initial mandates and advance their own policy preferences in ways their political superiors do not expect. Experts frequently act as agenda setters, drafting and implementing policies that affect the interests of political and social actors. Technical knowledge provides its bearer considerable power in the Latin American state, much more power than is recognized by other theoretical perspectives.

By using their knowledge to make public policy increasingly sophisticated, technocrats become essential to *políticos* and other socioeconomic actors who, without experts, could not manage the increasingly complex state. This gradual process itself reinforces the position of *técnicos* (Grindle 1977, 423; Haas 1982, 4; Centeno 1993, 316; Palacios 2001). But my findings also show that technocrats are not passive in this process. Experts frequently use their knowledge to purposively exclude less-technically trained actors from policy making, divert political and socioeconomic pressures, build informal and formal institutions that guarantee their positions' stability, and form alliances with other actors to advance their technical preferences. Through these strategies, experts actively aim to build technocratic autonomy and achieve continuity in their positions.

The strength of this autonomy, however, varies across policy sectors. In some policy areas experts enjoy high autonomy, successfully marginalizing other actors from policy making, advancing their preferences, and adopting informal or formal rules that make technical expertise a necessary requirement for recruitment into the state. Under these facilitating conditions, experts are also more likely to entrench their power; technocrats gradually raise the political costs of appointing nontechnical individuals, thereby informally constraining even new incumbents to appoint officials with similarly high technical credentials. Only considerable crises linked to the experts' performance or abrupt political changes can now reduce technical control over these areas. By contrast, in policy sectors where this autonomy is lower, technocrats are restrained from advancing their preferences and remain more vulnerable to clashes with politicians and

TABLE 1.1. *Technocratic Autonomy*

	Limiting External Intrusion	Advancement of Policy Preferences	Continuity/ Entrenchment
High technocratic autonomy	Significant capacity to resist/veto the intrusion of other actors over policy making, even including experts' political superiors.	Significant success in advancing technocrats' policy preferences, including projects not necessarily supported by their political superiors.	More likely. Politicians are informally constrained to maintain a high technical level in these areas. Incumbents may choose experts with different policy perspectives, but high technical credentials are a necessary condition for appointment.
Low technocratic autonomy	Vulnerable to clashes with politicians and socioeconomic actors.	Higher difficulties to advance experts' policy projects, especially when these are not supported by the incumbent.	Less Likely. Politicians have more latitude to dismiss experts, enhance political control, and appoint nontechnical cadres.

other actors. Furthermore, in these policy areas, the passage of time alone does not necessarily guarantee technical entrenchment. New incumbents, or new ministers, are not required to respect technical continuity. Table 1.1. presents the differences between instances of high and low autonomy.

I propose that four factors explain the variation in the strength of technical autonomy found across policy sectors, and over time, in the same policy sectors. The first two are related to what I refer to as "the politics of policy," while the second two are related to the technical aspects of policy. These first two factors carry most of the weight to explain this variation: they indicate which policy areas are more likely and less likely cases of technocratic autonomy. These two factors enhance the need for technical expertise within a policy area and crucially limit the incumbent's capacity to freely manipulate, or drastically reduce, technical expertise once established. The first two factors are as follows:

The Cost to the Incumbent of Poor Policy Performance – The most common threat to experts' autonomy in Latin America is the president's or ruling-party's interest in controlling the policy sector where experts are appointed. Although experts are often appointed to carry out politicians' policy goals, clashes inevitably arise, especially when a new executive takes office. A crucial test for technical autonomy is surviving these disagreements with political patrons.

What prevents politicians from firing experts, and why are they constrained to appointing only similarly technically trained cadres in certain policy areas? Part of the answer lies in the potential costs of intrusion. In policy areas where poor decisions can hurt the population and societal actors, creating negative political consequences, the costs of political meddling are prohibitively high; this is especially true if the negative consequences are evident in the short term, hurt a great number of citizens, affect high and middle classes, and are easily attributable to the political incumbent. On the other hand, the benefits to the incumbent for controlling a policy area are mainly increased access to patronage resources. Politicians are much more likely to forego patronage resources if the politicians themselves can be held responsible for the consequences of policy errors. However, if policy areas are of little interest to the public or societal actors, politicians face lower costs when replacing experts or intruding in existing technical areas. These political costs may increase over time because of the feedback effects of experts' policies or because of events that enhance public awareness of the consequences of poor policy, such as a political or economic crisis, thus contributing to technical continuity and entrenchment.

Plural and Balanced Constellation of Stakeholders[4] – Some types of policies affect the interests of multiple politically powerful, economic, and international actors, while others attract the attention of only a few stakeholders with limited political influence. If there is only one prominent actor with interests in a policy sector, this actor has significant power to block technical policies affecting its interests. But if there is competition between several powerful stakeholders, this balanced arrangement provides experts with more space to make their own choices. Ideally, stakeholders would like direct control of a policy area. When that is not possible, they prefer ceding control to technocrats rather than to competing actors, especially when those competing actors are politicians. These stakeholders see technocrats as a reasonable "second best" option that provides an informal check against intrusion by other actors. Competing stakeholders not only support technocrats as an insurance policy against other actors' policies, butexperts can also use these powerful actors as allies to advance technical preferences on those occasions when stakeholders' interests coincide with their own. This factor is also crucial to explain technocratic entrenchment. Building upon North's dictum that institutions gain stability "only if they are supported by organizations with an interest in their perpetuation" (North 1995, 25), I propose that powerful political and economic actors contribute to technical entrenchment. Over time, these actors come to

[4] As discussed more thoroughly in Chapter 3, I use the term "stakeholder" in a broader way than is usually used in public policy theory, which generally reserves the term for citizens, interest groups, or civil society actors affected by policies adopted in a policy sector. Here, in contrast, "stakeholders" refer to political, international, and socioeconomic actors whose interests are affected by policy making. IFIs or even the political opposition will also be considered stakeholders for my theory.

perceive experts as guarantors who maintain technical policy and predictable rules and thus have an even greater stake in the experts' continuity.[5] In areas where such stakeholders do not exist, technical entrenchment will be more difficult because the passage of time alone does not have similar positive feedback effects and experts remain vulnerable to clashes with politicians.

These two factors determine, to a considerable degree, the variation in technocratic autonomy within state agencies – in my examples, between high technocratic autonomy in economic policy and low autonomy in health policy. Nonetheless, two additional factors related to the technical aspects of policy also contribute to explain this variation. These two factors are not as important as the previous two in regard to the initial technical appointment or to limit political intrusion once technocrats emerge within a policy area. Nonetheless, when high, these factors enhance technical autonomy by further limiting political intrusion into policy areas. Thus, these factors help to explain why technocratic autonomy emerges and achieves continuity in less likely policy areas, and why it fails to do so in others where it would seem likely.

Policy Complexity – A third factor influencing variation in technical autonomy is the level of technical specialization required in order to understand policy options within a given issue. Policy sectors can be highly technical, or become so because of changes in the professional, academic, and policy communities. Policy complexity enhances autonomy by increasing both the information asymmetry between experts and politicians, and the uncertainty about the likely consequences of certain policies. Due to their lack of technical understanding, politicians rely heavily on experts to make these assessments, thereby reducing their own control over policy decisions.

Expert Consensus – This factor refers to the degree of agreement among experts on "correct" policy in a given area. In policy sectors where expert consensus about what constitutes adequate policy already exists, there is less space for political intrusion. Consensus enhances technical experts' autonomy by making technical preferences appear legitimate and reasonable while ruling out challenges from other technically trained individuals. When such a consensus exists, there may be discussion about policy details or priorities, but there is no debate among experts about the broad parameters of adequate policy. Technical consensus, then, provides experts with "an authoritative claim to policy-relevant knowledge within that domain or issue area" (Haas 1992, 3). Lacking fundamental disagreements among experts, it is more difficult for politicians to circumvent technical advice and policy prescriptions by replacing entire technical cadres with others holding preferences closer to their own. The lack of such divides forces politicians to accept the experts' views.

In the following chapters, I show how these factors determine in which policy areas experts are more likely to acquire technical autonomy and maintain it

[5] About feedback effects, see Pierson 1994, 42–50; 2000.

over time, and where technocrats are likely to remain weak and vulnerable to political intrusion. These factors explain the varied political fates of the three groups of economic experts (Colombia 1958–2013; Peru 1980–1983; Peru 1990–2013) and the three groups of health experts (Colombia 1966–1970; Colombia 1993–2013; Peru 1990–2002) studied in this book, which are summarized in Table 1.2.

In two of the three instances of economic policy making, Colombia 1958–2013 and Peru 1990–2013, all of these favorable conditions were present. Both of these cases show high levels of technical autonomy and significant continuity over time. A third economic episode (Peru 1980–1983) provides intermediate findings in a "more likely" area for technical autonomy. An unbalanced constellation of stakeholders and a lack of awareness of the political costs of profligate economic policy, combined with a lack of technical consensus, explain the fall of MEF experts in this earlier episode of economic technocracy.

The three health policy episodes show a substantial degree of autonomy for a sector that was traditionally technically weak. Reform teams were successful in implementing significant changes in the ministry. But the unfavorable conditions related to health policy explain why in two of my three health policy cases – Colombia 1966–1970 and Peru 1990–2002 – technical autonomy eventually declined and, despite remaining at intermediate levels in the third case, Colombia 1993–2013, continues to experience more challenges than in economic areas. In this last episode, experts gained significant leverage during an ambitious process of reform that radically transformed the Colombian health system. The factors in my framework, particularly the costs of policy failure and those related to expert knowledge, help explain this intermediate outcome. First, creating a single national health system that included both the formally employed and poor citizens raised the costs of poor health policy outcomes for politicians. Also, due to the privatization of health care provision, new stakeholders emerged in the health sector that balanced the power of the Executive. Finally, health policy became more technical than it was in the past because of the spread of the health economics approach (discussed in Chapter 6).

There are also limits to this technocratic power. To demonstrate how the influence of well-entrenched technical policy makers can fade, I briefly discuss in the conclusion cases from Argentina, Bolivia, and Venezuela where this occurred. In these instances, not only did new policy makers enter the government bringing new priorities, but the overall level of technical power declined in these sectors. New appointees were more dependent on politicians. This occurred when a crisis delegitimized technical policy recipes, thus reducing the political cost of firing experts and/or when an abrupt change of political power provided politicians with a strong popular mandate that allowed them to drastically reduce the experts' power. The factors I highlight help to explain the rise, stabilization, and fall of technocrats in the region.

TABLE 1.2. *Episodes and Outcomes*

Episodes Factors	Economy: Colombia (1958–2013)	Economy: Peru (1990–2013)	Economy: Peru (1980–1983)	Health: Colombia (1993–2013)	Health: Colombia (1966–1970)	Health: Peru (1990–2002)
Politics of policy						
Cost of bad policy performance	High	High	Medium/high	Low first, then medium due to policy feedback effects	Low	Low
Constellation of stakeholders	Plural and balanced	Plural and balanced	Plural and unbalanced	Dominant actor but more plural after the reforms	Dominant actor	Dominant actor, but balanced by IFIs during the 1990s
Technical aspects of policy						
Complexity	High	High	High	Medium	Low	Medium
Expert consensus	High	High	Low	Low	Low	Low
Outcome	**High autonomy, technical entrenchment**	**High autonomy, technical entrenchment.**	**Medium autonomy, no entrenchment.**	**Medium autonomy, meaningful continuity.**	**Low autonomy, no entrenchment.**	**Medium/ low autonomy, no entrenchment.**

RESEARCH DESIGN

To understand the autonomy of technocrats, I compare the implications of instrumentalist explanations and those of my own theory against evidence from economic and health agencies in Colombia from 1958 to 2013 and in Peru from 1980 to 2013. Despite some shared background conditions, Peru and Colombia are rather different cases within the region in regard to several factors that could, in principle, impact the emergence, strength, and autonomy of a "technocracy." For instance, *técnicos* arose in these countries in different time periods with distinct economic and political conditions, such as a deep economic crisis in Peru and a strong party system in Colombia. If similar findings apply across such relatively different settings, we can assume that they speak to other Latin American countries more broadly.

The understudied case of the Colombian economic technocracy is well suited for assessing technical power. Colombian technocrats emerged in the late 1950s when there was neither a severe hyperinflationary crisis nor dominant policy paradigm, such as neoliberalism, in the region. Thus we have to look beyond conjectural causes to explain the emergence of this technocracy. Further, we cannot attribute this strength to relatively weak political actors, as Colombian technocrats coexisted with well-organized parties for more than forty years. As the Colombian example shows, technical autonomy can live alongside strong and clientelistic patronage-based political parties.[6]

In each of the six episodes I explore, I document how and why experts' reform teams reached their positions within the state. Then I demonstrate experts' autonomy by highlighting how their preferences differed from those of other powerful actors, sometimes including their own political superiors. To further substantiate these claims, in each episode I discuss technical policy initiatives and their implementation, as well as the experts' veto power over other actors' initiatives. Finally, I show how the identified factors operated in order to explain the outcomes in each episode.

My focus is on executive agencies in which experts, in formal terms, depend on politicians in order to keep their positions. The reason I have not included formally insulated institutions where expertise is also found, such as central banks or regulatory agencies, is the difficulty in determining how much these formal institutional protections account for technical experts' autonomy. In this respect, I am focusing on "least likely cases" (in formal institutional terms) for gaining technical autonomy. If technocrats are autonomous where they lack formal protections, it is safe to assume that they are more likely to have such

[6] Clientelistic linkages between politicians and citizens "are based on direct material inducements targeted to individuals and small groups of citizens whom politicians know to be highly responsive to such side-payments and willing to surrender their vote for the right price" (Kitschelt and Wilkinson 2007, 2). In this work I'll interchange the terms clientelism and patronage as synonymous.

autonomy in areas dealing with similar policy issues that do enjoy institutional protections.

The perceived durability of the Peruvian and Colombian economic technocracies may create the impression that the continuity seen within these technical areas is above the influence of changes in political or economic factors. To show the limits of even strong technocracies in the region, in the concluding chapter I review three shadow cases – Argentina, Bolivia, and Venezuela – where established technocrats lost power. I illustrate how abrupt changes in the two politics of policy factors identified in my theory explain the experts' demise in these instances.

This book addresses two common problems in previous studies of technocrats in Latin America. First, this work presents a cross-temporal analysis of technocratic involvement in policy making. Most previous analyses of experts' actions within policy sectors employ a "static approach" (Centeno 1993, 316).[7] These investigations resemble a snapshot – an image – of a given moment that does not document variation over time. Their conclusions, then, may only apply to that precise moment. A political-incentives perspective, for example, can dismiss brief spells of technical autonomy under crisis conditions as exceptions to "politics as usual." Consequently, some perspective across time, as the one offered in the following pages, is important for properly assessing competing instrumentalist theories as well as my own.

Second, by comparing two policy sectors, economy and health, this work also differs from previous studies that have focused almost exclusively on economic experts (P. Silva 1991; Conaghan and Malloy 1994; Centeno 1994; Teichman 1997; Centeno and Silva 1998; Conaghan 1998; Santiso and Whitehead 2006). This narrow focus seems inadequate if we want to develop a more general picture of experts' roles in contemporary Latin America. As mentioned, I propose that there are particular social and economic conditions surrounding economic policy that make certain policy areas "more likely" cases for technocratic autonomy. Contrasting economic and health policy allows me to precisely highlight these differences. Health officials are expected to have specialized knowledge in public health in order to face existing and forthcoming challenges, and to guarantee an efficient allocation of resources. Thus, health is a policy area in which technical expertise should also be a crucial criterion for appointment but where, however, technical autonomy is much less likely to emerge and entrench than it is in economic sectors. Comparing policy areas within the state allows for an exploration of the causes that explain the variation in technocratic autonomy across these areas.

Finally, my work draws on three sources of information. First, I conducted interviews with *técnicos*, politicians, IFI officials, business leaders, and civil society actors. A total of 142 individuals were interviewed: 98 in Peru and 44

[7] An exception is P. Silva 2008. Silva's book is a long-term and detailed analysis of technical policy making in Chile's history.

in Colombia. These interviewees include 26 individuals with current or past ministerial-level positions, such as ministers of finance or health (four were appointed ministers after being interviewed). In Colombia, I had the good fortune to interview former Presidents César Gaviria (1990–1994) and Ernesto Samper (1994–1998). Second, I reviewed official policy documents and press archives regarding relevant events involving experts and the other sociopolitical actors in both countries. Policy projects, political statements, and press declarations helped to document the preferences of these diverse actors. Third, I collected biographical information on cabinet-level economic officials from official archives and other sources in order to document the experts' career paths. Although this information was incomplete in some cases, the biographies show clear trends that allow for valid generalizations.

A comparative historical analysis of the work of experts in two policy sectors fills an important gap in the study of Latin American politics. It has been observed that "given their importance, it is surprising how little has been written in Latin America about experts and their political roles" and also that "no systemic and explicitly comparative analysis of these actors has been done" (Centeno and Silva 1998, 1–2). Through a comparative historical investigation of Colombia and Peru across two policy spheres, I assess the soundness of previous findings about experts in the region, and propose new, more robust results and arguments.

DEFINITIONS: TECHNOCRATS AND DEMOCRACY

"Who are technocrats?" has long been a disputed question in comparative politics, because they cannot be defined by easily discernible criteria such as level of education, area of expertise, position within the state, lack of political ties, etc. (Grindle 1977; Centeno 1993). There is broad agreement among academics, however, that what distinguishes experts from other actors, such as politicians or bureaucrats, is their higher level of expertise (Centeno and Silva 1998, 2; Teichman 1997, 32), specifically in applied sciences such as economics, engineering, agronomy, or health policy (Putnam 1977, 384).

David Collier's classic definition of technocrats captures this distinctive characteristic: "Individuals with a high level of specialized academic training which serves as a principal criterion on the basis of which they are selected to occupy key decision-making or advisory roles in large, complex organizations – both public and private" (Collier 1979, 403).[8] This criterion of expertise not only distinguishes experts from politicians and most bureaucrats, but also includes

[8] Similarly, other definitions highlight expertise as the main criterion of selection: "personnel who use their claim to knowledge (as opposed to representation or authoritarian control) to affirm their right to rule" (Centeno and Silva 1998, 2); "highly educated bureaucrats, most commonly with graduate degrees in such fields as economics or public administration in U.S. Universities" (Teichman 1997, 32); or "highly educated, professionally trained policy makers" (Williams 2006, 119).

as experts lower-ranking officials with high technical skills (e.g., advisers and consultants). Diverging from Collier, however, I refer only to individuals who occupy public decision-making or advisory positions and not to individuals in the private sector. Out of the 142 individuals interviewed in both countries, 78 shared these characteristics.

In regard to the political regime type in which contemporary technocrats operate, I adopt a procedural concept of democracy. Following studies on democratic transitions, such a definition disaggregates regime type from regime quality (e.g., excluding "quality of democracy" measures like high electoral participation, effective socioeconomic rights, etc.). However, the definition has to be broad enough to include the effective respect for political rights that allow for meaningful competition for power. Democracy, then, is defined as "a set of institutions that, in the context of guarantees of political freedoms, permits the entire adult population to choose their leading decision-makers in competitive, honest, regularly scheduled elections and to advance their interests and ideas through peaceful individual or collective action" (Weyland 1996, 8).

According to this definition, both Colombia and Peru were democratic for a significant majority of the periods under study, allowing for a proper assessment of experts' role in democracies. At the same time, both countries experienced episodes of either restricted democracy or, in the case of Peru, authoritarian government, which allows for an examination of technical power under non-democratic regimes as well.

Colombia's National Front political pact restricted electoral competition for sixteen years between 1958 and 1974. The two traditional parties agreed to run a single candidate, alternating between parties, while all other elected positions were shared equally. Nonetheless, third parties could compete against the Front's candidate if they registered themselves as "liberal" or "conservative," even if they had no real attachments to these parties. While the National Front period fails to meet our procedural definition of democracy due to its limited contestation, Colombian politics during the time remained quite competitive, especially when compared to other Latin American countries during those years (Wilde 1978). Peru, on the other hand, was democratic from 1980 to April 1992, when Fujimori's coup suspended democracy until new elections were held in November of that year. Later in the decade (1997–2000), the regime restricted some liberties in order to gain reelection, eroding the democratic standard. Since 2000, when Fujimori fled office because of a corruption scandal, democracy in Peru has remained competitive.

THEORETICAL RELEVANCE: POWER, EXPERTISE, AND DEMOCRACY

The present study of technocratic roles and modes of operation in Latin America has significant implications for comparative politics. First, my evaluation of the power of social and political actors in Latin America suggests limits on existing instrumentalist explanations for the role of technocrats. While these theories propose that technocratic power results from one dominant actor – politicians, IFIs, or business interests – that uses experts as a tool to implement their preferred policies, my findings show that experts frequently act autonomously, trumping the interests of influential political and economic actors.

These results demonstrate the importance of certain state actors for the analysis of politics in Latin America (Weyland 2002a, 75). Bureaucratic weakness has led some to theorize the state as a playing field for political or economic forces. This image is misleading. As this work shows, some state actors, such as technocrats, can quite frequently constrain and affect the outcome of competition between these political and socioeconomic forces.

At the same time, by analyzing different policy areas, the book systematically explores the different factors that condition experts' power across the state. Looking closely at the interaction of technocrats with other political and economic actors allows me to specify the mechanisms that make expertise an effective source of political influence in Latin American democracies, its variation across policy areas, and the limits of such political resource. In sum, "technocratic autonomy" constitutes a valuable concept for the study of politics in the region.

Second, and closely related to the previous point, studying the role of *técnicos* as policy makers calls attention to the "output" side of politics, an often neglected topic in political science. While studies of the "input" side of politics examining parties, civil society, and social movements in Latin America have been common in recent decades, there have been fewer investigations of political outcomes. The "output" side of politics deserves more attention because it allows us to assess the power of sociopolitical actors (Carpenter 2001, 11; Weyland 2002b, 66; 2006, 11–14). Furthermore, as this book illustrates, policy output can shape politics in dramatic ways.

A third contribution of my analysis is that it helps explain the oft-noted variation of capacity within Latin American states, with some areas resembling professional sectors of developed states while others remain weak and ineffective. My theory provides insights about the likelihood of technical agencies establishing their authority, and why this likelihood is higher in policy areas such as economics than in others such as health. As Guillermo O'Donnell wrote almost two decades ago, "It says a lot about actual policy priorities that, while in several countries efforts have been recently made to enhance the economic policy-making area of the state, except for Chile, to my knowledge, no effort has been made in relation to the social policy area of the state" (O'Donnell 1996, 9). Although O'Donnell's words are not as true today as they were when first enunciated, this point still remains relevant for several Latin American countries. By looking at different policy areas and their dissimilar relevance for political and economic actors, the following pages delve into the reasons behind these differential priorities.

Finally, my work has implications for debates about the quality of Latin American democracies in which technocratic influence is often regarded as detrimental (Centeno 1993; Conaghan and Malloy 1994; Teichman 1997; Centeno and Silva 1998). The argument is that experts may tie the hands of even reformist politicians by undemocratically limiting the set of policy options. Therefore, public policy is no longer a matter of electoral choice, but decided by the preferences of technical elites or the powerful actors they allegedly represent. By assessing the ability of technocrats to trump political preferences under democracy, I focus on one of the most disputed issues in political science and democratic theory: the power of non-elected technical experts in democratic societies.

Diverse authors highlight occasions when technical bureaucracies in modern societies achieve the power to implement their own preferences even against the will of their democratically elected political superiors (Suleiman 1974; Carpenter 2001; C. Johnson 1995). The authority of experts in contemporary societies has captured the attention of a variety of social and political theorists (Bell 1973; Freidson 1984; Sarfatti Larson 1984; Halliday 1987). Ideal-type definitions of technocracy and democracy as political regimes are in direct opposition. A technocracy is defined as "the administrative and political domination of a society by a state elite and allied institutions that seek to impose a single, exclusive policy paradigm based on the application of instrumentally rational techniques" (Centeno 1993, 314). A democracy, on the contrary, has as one of its main characteristics the giving of equal weight to citizens' preferences, even if these preferences oppose "efficient" policies. A technocracy, then, is an exaggerated version of the dreaded "bureaucratic state" that has haunted political theorists since the nineteenth century (Tocqueville 2000; Weber 1946; Bell 1973).

This book does not claim that "rule by experts" in the literal sense exists in Colombia and Peru, or elsewhere in Latin America. As we will see, experts coexist with other powerful actors, and they do not always win their battles. Under some conditions, even powerful technocrats, well entrenched in the state, can lose their positions. My main point is that experts are strong actors, with significant autonomy and influence in the region. And in some policy areas, such as economic policy making in Colombia and Peru, experts can achieve strong continuity, pull elected politicians toward their preferences, and even run government to a considerable degree, making the aforementioned democratic tension undeniable.

There is an extensive body of literature highlighting the undemocratic tendencies of technocrats in Latin America, particularly their secretive and unaccountable policy style, as well as the controversial content of their sponsored policies. While this book adds evidence to some of these charges, it also contends that technocratic actors may have some positive effects on developing democracies, such as restraining political appetites, enhancing state capacity, initiating broad reforms in crucial areas of little interest for politicians, and even making public policy more responsive to citizens' needs (Williams 2006). By reducing politicians' patronage-oriented control over state resources and public policy, technocrats are also exercising a salutary effect on democracy. In the absence of professional public service, experts' power to restrain politicians may resemble the positive influence exercised by bureaucracies over politicians in developed democracies. Paradoxically, then, non-elected experts have also contributed to the survival of democracy for three decades by making these regimes more stable and attuned to citizens' needs.

ORGANIZATION OF THE BOOK

To develop my argument, I start with a discussion in Chapter 2 of the limits of the three instrumentalist perspectives on the experts' role in Latin American democracies. Chapter 3 proposes a theory of technocratic autonomy that explains the source of the experts' autonomy and why it varies in strength across policy sectors.

Chapter 4 analyzes the emergence and entrenchment of economic technocrats in Colombia from 1958 to 2013. Since the late 1950s in Colombia, self-reproducing groups of experts in the Ministry of Finance (MF) and the National Planning Department (NPD) have been in charge of key aspects of economic policy. These technocratic elite have helped Colombia, which had been trapped in a spiral of drug trafficking and political violence, to maintain a record of stable, prudent, and creative economic policy (Urrutia 1991; Thorp and Durand 1997; Botero 2005; Hershberg 2006).

Chapter 5 discusses the tenure of economic technocrats in Peru since 1980. The early 1980s shows that technocrats' failure to achieve continuity in the Ministry of Economics and Finance (1980–1983) was due to the opposition of political and business sectors. Peru witnessed the emergence of economic experts in the Ministerio de Economía y Finanzas (MEF) during President Alberto Fujimori's government (1990–2000); these *técnicos* oversaw the implementation of drastic market reforms in response to the country's severe economic crisis (Wise 2003, ch.6; Cortázar 2006). Technocrats maintained autonomy after the crisis receded and through the successive governments of Valentín Paniagua (2000–2001), Alejandro Toledo (2001–2006), Alan García (2006–2011) and Ollanta Humala (2011–). In 1990, technical expertise was enhanced within the MEF and has entrenched itself since then.

Chapter 6 focuses on two periods of health expertise in Colombia: the health reforms under the National Front, especially under President Carlos Lleras from 1966 to 1970, and the ambitious process of health reform during the 1990s. In the first Colombian case, experts played a leading role in the government and built a strong Ministry of Health during the National Front years (1958–1974), especially under the presidency of Carlos Lleras Restrepo (1966–1970). Nevertheless, these experts failed to maintain control of the ministry once the National Front ended and political parties captured the health system to strengthen their patronage-based networks across the country. In contrast, the second Colombian health policy episode (1993–2013) represents a "least likely" case where technical autonomy is meaningful and continuity significant. It is too soon to declare technical entrenchment, but there is significant stability, and technical ministerial picks are now common.

Chapter 7 analyzes the work of experts in Peru's Ministry of Health from 1990 to 2002. The International Cooperation Office in the Peruvian Ministry of Health concentrated high levels of expertise from 1993 to 2002. During these years, experts oversaw broad policy reforms both to rebuild and restructure the health system after Peru's internal war against the Shining Path (1980–1994). This higher level of autonomy is explained by medium scores in two of my factors. The growing interest of IFIs in health reform created a more balanced constellation of stakeholders. Moreover, health policy became more complex through the emergence of health economics, a policy approach that became prevalent in the 1990s. Starting in 1999, however, experts' influence gradually weakened because of resource shortages and clashes with political interests, as well as a decline in IFIs' interest in health reform. These changes led to the demise of experts in 2002.

Chapter 8 concludes by exploring the generalizability of my theory of technical autonomy. It discusses cases of technocratic downfall in Argentina, Bolivia, and Venezuela and also presents the broader implications of my findings.

2

Technocrats in Latin American Democracies: Agents or Actors?

In a process starting in the late 1980s, and deepening with the neoliberal reforms of the 1990s, technocrats have become important actors in Latin American democracies. Experts' heightened role under democratic regimes has stimulated a lively theoretical debate about their influence in policy making and their effective independence from other sociopolitical forces. Three instrumentalist theories, each derived from a broader theoretical perspective, provide competing explanations for the heightened role of technocrats in contemporary Latin America and, more particularly, the question of how officials who lack tenure or institutional protections can become influential political actors. These theories find the source of experts' influence in other actors; experts are the agents of these powerful principals.

By contrast, my perspective highlights experts' autonomy: technocrats are independent actors who contest regional politics in their own right. I argue that experts "formulate and pursue goals that are not simply reflective of the demands or interests of social groups, classes, or society," thus fulfilling the most widespread definition of autonomy in comparative politics (Skocpol 1985, 9). Experts not only have distinguishable preferences from social groups but as I show, also enjoy a considerable degree of autonomy from their political superiors.

In this chapter, I present the three instrumentalist perspectives and discuss their shortcomings in capturing the role of *técnicos* in the region. Although they provide important insights into understanding technocratic power and may be empirically accurate in particular cases, my evidence shows they greatly overestimate the leverage of particular powerful actors over experts and fail to capture the autonomy experts have in the region. In the next chapter, I develop a theory of technocratic autonomy to explain these findings, including the variation in the strength of technical autonomy both across policy sectors and across time in the same policy sector.

Before proceeding, it is worth noting that both the three instrumentalist theories and my technocratic autonomy model all reject a functional perspective that explains the greater importance of technocrats as a natural response to the complexities of state management caused by modernization and globalization (Williams 2006, 119). From a functional perspective, experts became relevant actors simply due to Latin American states' need to respond to the challenges of a competitive global environment and increasingly technical domestic policy; states employ experts to fulfill the need for greater technical knowledge and will only fail to do so if such expertise is unavailable. A functional perspective certainly helps explain the need for expertise in the region, because countries that must deal with global integration, international demands, and complex policy issues have a big incentive to increase their expertise. And, clearly, expert management will not be feasible if states lack educated professional cadres from which they can recruit this expertise. However, if this functional perspective were correct, we would expect to see experts having approximately equal levels of importance and influence in countries possessing similar technological and human resources and facing comparable external pressures. What we find in Latin America are different levels of technical expertise, both across countries and within countries over time. The increasingly complex policy environment may create demand for more technical expertise in certain policy areas, but how different countries respond to this need is a political question that remains unexplored (Schneider 1998, 79). The four competing theories, therefore, all reject this functional perspective.

POLITICAL AGENTS

Electoral-incentives theories claim that experts derive their power from politicians. Such a political-agents perspective assumes that politicians usually want to maintain control over state resources, which are valuable for advancing their electoral interests. However, this perspective predicts that under certain conditions, politicians restrain their control over patronage and delegate the power over policy making to experts (Geddes 1990, 218). The insulation politicians provide gives technocrats considerable protection from other socioeconomic actors and allows them to adopt technical policies. How can a theory based on instrumental political interests explain the "enlightened" choice to delegate power to technical experts?

This literature provides two answers to explain this delegation. On the one hand, politicians are aware that seeking shortsighted and narrow interests can produce inefficient outcomes that ultimately harm their longer-term electoral interests.[1] Thus, on some occasions, politicians "tie themselves to the mast"

[1] For a description of this approach to explain market reforms in the developing world, see Bates and Krueger 1993, 464–465.

to guard against their own irresponsible short-term behavior with the goal of maximizing their medium-term political payoffs. Delegating authority to *técnicos* is one way to do so, but at the cost of restricting their use of patronage resources. From this perspective, some types of crises that threaten their electoral chances (e.g., hyperinflation) should lead politicians to bolster technical management. A second explanation sees this delegation as a way of directly advancing incumbents' interests. Presidents or ruling parties may sometimes advance their political goals through broad reforms that require specialized knowledge; therefore, they appoint experts to carry out these reforms (Geddes 1995, 213; see also: Geddes 1990, 213).

Nevertheless, the key assumption of this perspective is that no matter how much power is delegated to experts, politicians remain the principals in this relationship. The power of appointment and dismissal gives politicians extensive influence, and ultimately control, over experts. This political control precludes experts from gaining autonomy: technocrats may temporarily enjoy considerable latitude of action, but their fate always remains dependent on political discretion.

Barbara Geddes, for example, argues that when the benefits of patronage are outweighed by the benefits politicians can gain from policy reforms, they create technocratic "pockets of efficiency" to enact these reforms (Geddes 1990; 1994). Instances of state autonomy are described by Geddes as politicians' insulation of state agencies (and the experts working in them) in order to advance their own political interests (Geddes 1994). Given the high political benefits of patronage in Latin America, Geddes expects instances of politicians insulating experts to remain rare, and when they do happen, to be short lived. Once the crisis has abated and the electoral calculus no longer favors policies promoted by experts, or when new incumbents with different preferences reach power, technocrats will be removed and technical areas will again succumb to patronage-oriented politics (Geddes 1990, 231–232).

Not all proponents of political-agents theories use such a restricted understanding of political interest as does Geddes. By attributing such a crucial role to patronage, Geddes underestimates the likelihood of technocratic appointment even within a political-agents perspective. Other theories suggest that if certain conditions (e.g., a severe economic crisis) make the provision of some collective good more important for electoral success than does patronage, the probability of technical recruitment increases. This is Beatriz Magaloni's argument for explaining the appointment of neoliberal technocrats in Mexico during the 1980s (Magaloni 2006, ch. 3). Magaloni argues that politicians' public commitment to drastically reduce patronage can be a crucial electoral asset during a severe economic crisis. Amid the debt crisis of the 1980s, Mexican voters could no longer be swayed with patronage expenditures because they had learned that irresponsible economic policy was bad for their long-term interests. At the same time, Mexican politicians learned that pursuing such irresponsible policies eventually had high electoral costs. Therefore, Institutional

Revolutionary Party (PRI) politicians appointed neoliberal technocrats as a signal to voters of their commitment to responsible economic policy. Just as technocrats in the 1950s and 1960s increased spending for their PRI bosses, neoliberal technocrats served their political masters by reducing inflation and public debt. Magaloni predicts that the end of the crisis should lead to the demise of these allegedly powerful neoliberal *técnicos*.

Although coming from a different theoretical perspective, Ben Ross Schneider explains technocrats' emergence in Latin America after the debt crisis in terms of politicians' urgent need to attract external investment to their countries. Afraid that economic turmoil would negatively affect their electoral interests, politicians appointed orthodox economic experts to "signal" their commitment to protecting foreign investments (Schneider 1998). As in previous theories, Schneider argues that experts remain dependent on politicians' preferences as they lack "independent, enduring sources of power" (Schneider 1998, 90). If economic conditions change, and/or external funds become less important for politicians, technocrats will lose their control over policy. As Schneider writes, "The term technocracy is a misnomer" (Schneider 1998, 90).

Under these political-agents perspectives, assessing whether autonomy is real or merely delegated is difficult. However, these theories have two crucial and testable implications: experts are expected not to (1) trump key political preferences, especially electoral interests, or (2) remain highly influential over long periods of time. No matter how much power politicians delegate, these theories presuppose that politicians will overrule technically based decisions if they clash with their electoral interests. Similarly, these theories suggest significant technical influence to be short-lived. When the policies preferred by *técnicos* are no longer popular or necessary, especially once crises end or new incumbents with different preferences enter office, this perspective predicts that technocrats will adopt the policy preferences of their political patrons or face dismissal.

My research shows that a "political agents" perspective fails to capture the role that experts play throughout the region, seriously underestimating the power of technocrats vis-à-vis political actors. Technocrats' expertise grants them a level of autonomy that cannot merely result from political delegation. The cases show experts acting as agenda setters, advancing reforms that frequently affect key interests of their formal political superiors. In addition, against the most basic assumption of electoral incentive theories, I document how politicians have sometimes lost effective control over policy making, giving *técnicos* considerable latitude to craft and implement policies that transcend political preferences.

As discussed in Chapters 4 and 5, economic experts in Colombia and Peru effectively restrained key political interests. Economic experts' influence in these countries allows them not only to resist political pressures but also to advance their own policy preferences. After they emerged in Colombia amid the political violence and fiscal crises of the 1950s and 1960s, and Peru during the economic catastrophe of the late 1980s, experts entrenched themselves in economic institutions. Highly educated appointees in Colombia's Ministry of Finance (MF) and National Planning Department (NPD) and in Peru's Ministry of Economics and Finance (MEF) have acted as effective checks on politicians by controlling the budget and restraining the public deficit. These cases show experts taking an active role in parliamentary politics, frequently opposing laws debated by Congress because they run counter to experts' policy preferences. Experts also challenge proposals from members of the executive branch (e.g., the president or ministers), showing that they are not merely agents of their appointing superiors. In both countries, I document how experts implemented reforms that affected electoral interests or restrained political prerogatives, such as the adoption of procedures designed to control expenditures and resource distribution. Contrary to a political-agents perspective, then, economic experts have had considerable leverage over politicians.

In both cases, this ability to resist political pressures and advance policy preferences remains high over time. There is some variation, especially when facing personalistic and politically experienced presidents, but technocratic independence from politicians has remained fairly stable in economic policy areas. Experts' control over public spending, for example, persisted during good economic times, when politicians eagerly demanded more resources for their constituencies. The fact that conflict over a key political interest – namely, patronage resources – occurred under strong economic conditions contradicts a political-agents perspective that ties experts' power to crisis conditions. As highlighted by previous works, economic experts' power enhances during crises (Weyland 2002a; Teichman 1997), but the saliency of technocrats remains high even after crises have passed. Technocratic influence over economic policy, then, cannot be described as an episodic phenomenon.

This autonomy is not only found in economic ministries, which are "more likely cases" for technical autonomy. Chapters 6 and 7 document how health experts achieved considerable control over policy and advanced reforms that clashed with political interests. In both countries, we find episodes in which experts designed and enacted policies that reduced political intrusion in the health system. Even if this autonomy remained more vulnerable, it certainly exceeded the expectations of a political-agents perspective. Further, the experience of the recent Colombian health reform (1991–2011) shows that technical autonomy can achieve significant continuity in the health sector.

My findings also point to a second crucial limitation of a political-agents perspective and, more broadly, of theoretical perspectives that place politicians as the cornerstone of Latin American politics. I document how several reforms that were implemented in Latin American states were not closely monitored by politicians. Rather than seeking to enhance their electoral interests through public policy, as expected by a political-agents theory, politicians often remained ignorant of technical procedures and the complex reforms taking place. Although these reforms frequently affected their interests, politicians were unaware of their full implications. These cases demonstrate that experts frequently escape the control of their alleged principals.

Due to the complex nature of economic policy decisions, this lack of awareness is very clear in the economic episodes I explore, but it is also quite evident in the health policy cases. As in prior works, my study also shows that health experts helped advance reforms in Latin America even in the absence of political involvement (Nelson 2004, 31–32; Ewig 2010). Peru's health reforms in the 1990s, for example, were enacted with very little political involvement; experts led reforms in which politicians, including even the minister of health, played limited roles (Arroyo 1999, 549–551; Ewig 2010, 98; Ugarte and Monje, 1999, 576). Most of these changes were approved through ministerial or presidential decrees, and the policies were rarely discussed outside of the technical teams and consultants hired to design them (Ewig 1999, 490–493; Interview with Meloni 2007).

Even if it fails to capture experts' role in Latin America, the political-agents perspective does help explain the initial appointment of technocrats. As discussed later, this perspective's proposed mechanisms for technical appointment are useful for my own theory. In several of the episodes I explore, experts secured appointments at the state level due to political decisions prompted by the need for specialized technical knowledge to solve sensitive issues. Politicians frequently turned to technocrats in the midst of a crisis or under pressure from external actors (e.g., when politicians had to adopt stabilization plans, achieve development goals promoted by external agencies, or comply with the requirements of conditional loans). Viable alternatives were not available, as parties and bureaucracies lacked the necessary expertise, and the costs of doing nothing were too high. In addition, my own theory adopts the cost-benefit evaluation highlighted by this approach to explain why technocratic autonomy is higher and continuity more likely in areas of government where poor policy performance can inflict high political costs on politicians.

In conclusion, a theoretical perspective based on short- and medium-term electoral interests cannot explain the political autonomy of economic and health experts. Experts do not unilaterally depend on politicians, but instead command considerable autonomy and influence – in some cases even after a technocrat's political usefulness has declined. Furthermore, political-agents perspectives overestimate politicians' control over the design and adoption of public policy, as well as their ability and willingness to monitor experts. Thus, while a political-agents perspective helps to account for the initial decision to empower technocrats, and technical autonomy is more likely in politically sensitive areas, it fails to recognize the high degree of technical autonomy experts achieve and the persistence of this autonomy over time. Technocrats are not simply instruments of political actors.

AGENTS OF INTERNATIONAL FINANCIAL INSTITUTIONS

If technocrats' power is not derived from politicians, does it come from the backing of powerful foreign economic institutions? An "external agents" perspective argues that technocrats gain independence from politicians by becoming informal agents of international financial institutions (IFIs). Although no author goes as far as to assert that particular experts are directly chosen by IFIs, the implication is that only experts who follow IFI policy agendas are acceptable to these institutions. IFIs may not have direct control, but their immense power over the state trumps other available technical choices. From this perspective, IFIs, such as the World Bank (WB) or the International Monetary Fund (IMF), exercise pressure over politicians to insulate technocrats they approve from political and economic interference (Stallings 1992; Kahler 1992; Van Dijck 1998). Technocrats are the domestic representatives of these external hegemonic forces; they are powerful as long as they follow IFIs' preferred policies.

This external-agents perspective has its roots in the dependency theory developed in the 1960s. This theory questioned the basic assumption that developing states are sovereign, arguing that they were in reality subordinated to the interests of developed capitalist countries. There was less agreement among these authors about the mechanisms of this external control and how much independence domestic actors could achieve from these forces (Frank 1966; Cardoso and Faletto 1979).[2] Softer versions of dependency theory give more importance to the state and domestic business coalitions, which are sometimes in conflict with external interests (Evans 1979; Mahoney and Rodríguez-Franco, forthcoming, 5–7). But in general, these theories expected external forces to have a strong, fundamental influence over middle-income Latin American states.[3]

An "external agents" perspective attributes to IFIs the power to shape domestic politics through various avenues (Kahler 1992; Stallings 1992). One way is to employ "linkages," defined as indirect modes of intervention and support. Linkages between IFIs and states, for example, include promoting policies and training local officials. IFIs can exercise their influence by recommending that states adopt policies that have either been viewed as successful in other countries or regions (e.g., cash transfers as a means to alleviate extreme poverty) or that are regarded as necessary to face current and future threats (e.g., programs for epidemic control). These linkages have a long history in Latin America. Economic missions of foreign experts promoting broad institutional reforms have been common in the region (Drake 1989; Sandilands 1990;

[2] For reviews of dependency theory, see Leys 1996, 11–19; Mahoney and Rodríguez-Franco, forthcoming.

[3] Although international businesses operating locally also constitute external economic forces, I include these business interests in the next section. External and internal business actors comprise what I call business interests.

Cepeda Ulloa and Mitchell 1980, 238–241). Local experts receive frequent invitations to attend courses and conferences organized by IFIs, especially the World Bank and the Inter-American Development Bank. In these workshops and conferences, IFI policy recommendations and handbooks are publicized to local experts (Weyland 2006). These linkages include the frequent recruitment of domestic experts to serve as IFI cadres.

In a globalized economy, however, a stronger mechanism to influence domestic policies arises from a state's need to obtain approval from IFIs in order to access new sources of international credit and loans, which are especially important during economic meltdowns. The need for fresh credit to fulfill international and internal obligations leads states to sign conditionality agreements that include broad policy reforms. Stallings defines this kind of direct intervention by IFIs as "leverage" to distinguish it from linkages (Stallings, 1992, 52–58). States are "forced" to reform in accordance with IFI priorities in order to assure their economic survival.

According to this external-agents structural approach, local technocrats guarantee states' compliance with IFIs' policy recipes. IFIs exercise their leverage and condition external loans on the appointment of experts sharing their policy preferences. IFIs will even come to the defense of these technocrats if political or business actors threaten their operation. In this way, IFIs guarantee that those in charge of implementing their development recipes remain aligned with their interests.

Are *técnicos* informal agents of IFIs? Two facts suggest the answer is no and that these theories overestimate the influence of IFIs. First, the relationship of subordination postulated by this theory does not exist. IFIs may pressure politicians in developing states to appoint experts, but subsequently have little influence over the experts themselves. Second, the theory cannot explain the maintenance of technical autonomy over time.

My research shows that experts are better described as equals to IFIs. Local experts are neither subordinates to IFIs nor always in agreement with them. Quite frequently, *técnicos* reject IMF or World Bank suggestions in favor of their own preferred policy options. Even under conditions of economic distress, when IFI influence is at its peak, local experts sometimes choose to disregard the advice of these institutions.

Economic technocrats and IFIs in Colombia have long had a difficult relationship. In fact, Colombian technocrats first emerged not because of IFI lobbying but rather due to national politicians who promoted experts within the state in order to counterbalance the expertise of IFIs during intense international negotiations. As one of the first members of the Colombian technocracy noted, "To change this balance of forces, it was necessary to increase our national capacity of analysis, give more knowledge to the technical cadres of the state, and strengthen the fundamental aspects of the country's macroeconomic management" (Botero 2005, 21).

In Peru, we find similar evidence that experts do not act as agents of IFIs. Members of Fujimori's economic team claim that they autonomously believed that market reforms were the best solution to the economic crisis, thus denying that these reforms were imposed on them by IFIs. Supporting this claim is evidence that experts pushed for more drastic reforms than those proposed by IFIs (Weyland 2002a). Economic crises may increase IFI leverage, but the recruitment experts and the programs these experts implement are not the product of IFI control.

The health policy episodes also demonstrate that local experts frequently choose their own policies in direct opposition to IFI prescriptions. IFIs' interest in social sector reform during the 1990s put pressure on politicians to appoint experts in these issue areas and provide them with necessary resources, but it did not translate into control over health experts (Ewig 2004; 2010; Weyland 2006). As discussed thoroughly in Chapters 6 and 7, IFIs' interest in social policy helped trigger health reform, but local cadres had extensive authority to design and conduct these reforms.

The Colombian health reform, for example, was a unique model that generally followed the health economics approach promoted by IFIs such as the World Bank during those years (World Bank 1993) but diverged significantly from the privatization models international organizations favored. Similarly, Peruvian health experts successfully resisted pressures from economic experts in the MEF who demanded the adoption of health reforms in line with the Chilean health reform model and World Bank recommendations. Colombian and Peruvian experts' policy preferences also included reforms recommended by the World Health Organization (WHO), the Pan American Health Organization (PAHO), and the United Nations Population Fund (UNFP), institutions that lack the kind of leverage over countries often attributed to IFIs.

Moreover, even the adoption of IFI-sponsored policies does not necessarily indicate subordination to external forces, possibly reflecting a more complex relationship that provides additional evidence against a political-agents' perspective. As two Peruvian experts reported, domestic economic technocrats asked IFI representatives to include pension reforms that Fujimori had opposed in the IFIs' conditionality agreements in order to force the president's hand (interviews with Du Bois 2007 and Ugarte 2006). This example shows how local experts can use IFIs to "enhance their bargaining leverage with domestic opponents or hesitant political leaders" (Weyland 2005, 272). Similarly, IFIs' demand that technocratic appointees meet some specialized education requirements is less a mechanism of IFI control than a convenient rule that allows local experts to protect their autonomy from politicians.

A second limitation of an external-agents theory is its inability to explain experts' persistent autonomy over time. My research shows how economic experts keep their power even during good economic times, when developing countries are less beholden to IFIs and can disregard many of these institutions' policy prescriptions. Nonetheless, in both countries, technical influence has remained fairly constant across time. Similarly, health experts in Colombia remained relevant after IFIs reduced their interest in social policy reform, such as health policy.

An external-agents perspective, then, does not capture the role of experts in Latin America. Nevertheless, as discussed in the following chapter, some of this theory's insights are useful to explain technical autonomy. IFIs can be an important force behind the initial appointment of experts. Additionally, in some policymaking issues, the involvement of these institutions helps to explain the variation of technical autonomy across policy areas. My research shows how IFIs can pressure politicians to appoint experts, with the initial technical appointments in economic agencies in Colombia and Peru coinciding with economic crises during which IFIs strongly lobbied both countries. Similarly, the appointment of experts in both countries' health sectors during the 1990s followed new recommendations from IFIs for health sector reforms that considerably increased the technical sophistication of health policy (Nelson 1999; Kaufman and Nelson 2004). And in often underfinanced social sectors, IFIs and other international cooperation agencies provide valuable resources that enhance experts' capacity to develop and implement policies. Finally, as discussed in the next chapter, the interest of these powerful actors contributes to maintaining technical autonomy over time.

In conclusion, a theory that depicts experts as the informal agents of IFIs cannot account for the autonomy of technical cadres in Colombia and Peru. First, the relationship of subordination stressed by this perspective does not exist. More broadly, the theory exaggerates the leverage of IFIs over middle-income countries, such as Colombia, Peru, and most Latin American states. Although IFI leverage can encourage politicians to appoint experts in the state, respect their autonomy, and contribute to their continuity, experts are clearly not IFI agents.

AGENTS OF BUSINESS

If experts' power does not come from IFIs, does it emerge from business interests? Domestic structural perspectives depict economic technocrats as the agents of national business interests or foreign business interests operating locally. These approaches propose that as long as experts act according to business interests, they receive informal yet strong protection from political interference. Thus, the source of allegedly "technical" and "objective" policies is dominant business associations and business leaders. Technocrats are the chosen tools of economic elites to protect their interests by controlling public policy (Estrada Álvarez 2005; Durand 2006; Schamis 1999). As stated by Bates and Krueger when describing this perspective, "the empowerment of the economic bureaucracy represents, then, an attempt to stabilize the fortunes and protect the political triumph of particular interests" (1993, 465).[4]

[4] For a similar perspective in which business groups effectively determine state policies in Latin America, see Frieden 1985.

Control over state actors becomes especially crucial for business interests under democracy because elites lose many channels to protect their interests that had existed during previous authoritarian regimes. In democratic environments, business actors are especially interested in keeping politicians or state actors from advancing redistributive reforms (Schamis 1999; P. Silva 1991). A common criticism of neoliberal technocrats in Latin America is precisely that experts are closely tied to business elites who benefit from market reforms and are allegedly the new dominant economic interests in the region (Ahumada 1996; Estrada Álvarez 2004; Durand 2006; Schamis 1999; Teichman 2002).

This perspective has two versions: a strong structural version and a weak structural one. Strong structural perspectives see economic forces, especially business interests, as the primary drivers of politics in general, leaving little if any space for technical autonomy or state autonomy more generally. Examples of a strong structural perspective are presented by Schamis (1999) and Estrada Álvarez (2004; 2005). For Schamis, the influence of "insulated" state experts in Pinochet's Chile did not come from experts themselves but from powerful business coalitions (Schamis 1999, 237–238).[5] Similarly, what others see as a professional technocracy with autonomous preferences in the Colombian state, Estrada claims is a front for the interests of economic elites. Evidence of this dependence is emergence of new dominant business coalitions prior to the shift in technocrats' policy preferences.

Interestingly, Estrada's strong structural approach has space for some, albeit limited and relative, autonomy for state actors. A la Marx in his *Eighteenth Brumaire*, Estrada argues that during the 1970s, different sets of private interests in Colombia "balanced" each other out, giving the state leverage to act "independently." However, such an opening is expected to be short-lived, as one business sector will eventually gain dominance over the economy. In the Colombian case, Estrada claims, this period ended when financial and commercial interests strengthened in the 1980s (Estrada Álvarez 2005, 275).

Authors advancing a weak structural perspective do not share the assumption that state preferences always mirror structural forces. Less deterministic than strong structural perspectives, these views argue that state actors, such as experts, can have considerable leverage over business interests. Nonetheless, these authors argue that neoliberal technocrats are subordinated to the interests of business groups that benefited from market reforms. Because these reforms reduced the regulatory powers of Latin American states and favored certain economic activities over others, they created "winners" who could impose their policy preferences on the weakened state. Arce, for example, argues that Peru's market reforms were conducted by experts with significant autonomy from societal interests but that the winners from these reforms then quickly gained

[5] Even if less deterministic than Schamis, Eduardo Silva also depicts distributional coalitions of business sectors as an important cause of technocrats' power in Pinochet's Chile (E. Silva 1993).

leverage over economic policy agencies (Arce 2006). Similarly, Durand (2005) proposes that the winners of market reforms in Peru dictate economic policy, with financial, export, and mining interests in Peru controlling technocratic decision making. And although she leaves more ambiguity about the degree of dependence from business actors, Teichman also contends that economic technocrats in Argentina and Mexico are strongly linked to and influenced by the winners of market reforms (Teichman 2002).

My research confirms that there is significant contact between *técnicos* and business interests, especially since economic technocrats almost always have private sector work experience before or after their appointment. Economic policy making in Colombia, for example, has traditionally included consultations from business associations, especially the National Association of Coffee Growers (ANC) and the industrialist association (ANDI) (Schneider 2004, 128–150; Juárez 1993a; 1993b). Similarly, CONFIEP, Peru's all-encompassing business association, maintains a close relationship with the MEF. Experts not only often hail from the private sector but commonly leave their state positions to become corporate executives or chair business associations. In 2008, for instance, three influential Colombian business associations were chaired by ex-finance ministers or NPD directors. In Peru, this revolving door also exists: of ten economic experts I interviewed in 2007 who had recently left high-ranking state positions, eight were working for the private sector. Also, two of Fujimori's economic ministers were formerly presidents of business associations, while most of Alan García's and Ollanta Humala's economic ministers came from private financial institutions.

Although business actors are crucial to understanding technocrats' power in the region, and business does have significantly more access to experts than do other social groups, this perspective, in both its strong and weak versions, fails to capture the role of experts in Latin American democracies. Even if business actors are powerful in the region, especially where traditional countervailing forces like unions and leftist parties have lost power, this approach exaggerates the control of business interests over experts and the state more generally. I discard this theory for two reasons. First, even if a close relationship between economic experts and the business sector exists, experts' decisions do not betray dependence on private interests. Second, the health policy cases show that powerful experts can also emerge in policy sectors not dealing directly with business interests, though these technocrats remain more vulnerable to political inference than those in economic policy areas. These findings undermine strong structural claims that attribute experts' power to private interests.

First, the alleged subordination of economic experts to business interests is not confirmed by my research. All these perspectives assume experts' proximity to the business sector is a sign that technocrats are dependent on dominant economic actors in Latin America. But proximity between business and experts is not sufficient evidence of technical subordination. Comparative studies of bureaucratic independence in developed countries consider careers in business as evidence of dependence on private interests (Schneider 1993). In developed countries, such an assumption seems adequate, as the state, think tanks, and academia can provide similar working conditions as private enterprises. In Latin America and other developing countries, however, this standard is inadequate because the private sector may be the only source of qualified individuals for state positions and the only form of employment for technocrats after they leave the state. The technocratic phenomenon is explained partly by the absence of a strong and professional bureaucracy from where to look for experts to deal with complex policy issues: politicians and state actors find such knowledge in the private sector. In assessing experts' autonomy, the focus should not be just on their career paths, but also on the role played by experts while in office. Evidence from both countries demonstrates that past career experience and future career expectations do not determine economic officials' actions as assumed by this theoretical perspective.

Economic technocrats' decision making indeed shows that they enjoy significant independence from business interests. Experts have broad powers to conduct policy reforms in ways that affect seemingly dominant business interests. In line with previous works on economic policy making and market reforms, I do not find a strong relationship between business interest-group lobbying and experts' appointment or the initiation of economic reforms (Bates and Krueger 1993, 455; Geddes 1995). Moreover, experts' policies sometimes cause abrupt shifts in business activities and privilege some economic interests over others, resulting in enormous consequences for the structure of the economy, reducing the size of some sectors while allowing others to emerge. Divergence between economic policy and private sector preferences usually results from experts' need to respond to emerging international challenges, leading to policies that clash with business and political interests (Malloy 1992, 146–147). These conflicts show that while in office, experts are aware that broader interests and international threats often require the sacrifice of particular business interests. Even if economic experts have, or plan to have, stints in the private sector, they tend to advance more than private interests alone while holding office.

As discussed in more detail in Chapters 4 and 5, what experts perceived as economic threats led them to make drastic policy decisions that affected powerful business interests. In the 1960s, Colombian economic experts enacted a series of policies to make the economy more competitive and less vulnerable to external shocks but which put them at odds with powerful coffee and industrial interests (Mares 1993; Thorp and Durand 1997). A second case concerns the 1985 unexpected devaluation of the Colombian peso, which not only harmed business interests directly (Thorp and Durand 1997, 223) but relied on businesses being unaware of the changes about to occur. Similarly, market reforms from 1990 to 1991 were led by technocrats and not by business preferences. Interestingly, in some of these episodes, technocrats adopted policies that strongly deviated from the positions they lobbied for while chairing business associations (Urrutia 1983, 42–43). These examples show that their roles as government officials often trump their loyalty to business.

Market reforms in Peru and export-oriented reforms in Colombia show how experts' decisions can drastically alter businesses' power. Peru's market reforms, enacted in the early 1990s against the wishes of business interests, such as the industrial lobby, were still strong. The business groups that benefited from these reforms were not dominant when reforms were initiated nor sufficiently organized to lobby for deep market reforms. As a result of the reforms, industrialists lost considerable economic clout and political power, while other sectors saw their importance increase. Similarly, technical reforms in Colombia to move the country toward a more export-oriented development model in the 1970s and 1990s affected the power of the well-organized industrial sector.

Experts, however, do not always win these clashes. Just a decade before Fujimori's market reforms in Peru, during Fernando Belaúnde's government (1980–1985), an informal alliance of business, labor, and politicians defeated Peruvian economic experts trying to implement broad neoliberal reforms (Conaghan and Malloy 1994; Malloy 1992). Similarly, during the 1970s in Colombia, a coalition of business groups, especially industrialists, successfully lobbied for the dismissal of Finance Minister Rodrigo Botero. Botero's replacement was less orthodox than his predecessor and reversed some of the ongoing reforms. But these discrepancies show that experts' policies cannot be equated with business interests.

In addition, experts in Colombia have consistently sought to insulate economic policy making from business and political interests, a point that serves as further evidence against a business-agent perspective. They have managed to progressively reduce the influence of the private sector and politicians in policy making. In the 1960s under Carlos Lleras, *técnicos* adopted reforms to reduce congressional control over economic policy. They were worried that congressional involvement would open the door for strong lobbies from regional business interests. In the 1980s, Colombian economists sought to insulate the Monetary Board from private interests, because it not only determined inflation policy but also could provide special credit to business. Technocrats got their way in the 1991 Constitution, which assigns these responsibilities to an independent board of directors in the Central Bank. Rather than enhancing businesses' influence over public policy, Colombian *técnicos* pushed to reduce the opportunities for lobbying.

In Peru, there has been less conflict between MEF and the "winners" of the market reforms – especially financial, mining, and some export interests – which now have considerably more power than do industrialists. A good deal of consensus exists regarding policy between these newly dominant business sectors and experts in charge of economic policy (Arce 2006; Durand 2006). Certainly, such consensus is expected because economic reforms benefited certain economic activities in accordance with the reformers' preferences, thereby showing the triumph of experts over those business sectors that opposed their reforms.

While Peruvian experts are undoubtedly closer to the private sector than are their Colombian counterparts, as discussed in Chapter 5, it is overreaching to call such proximity subordination. Recently, Peruvian technocrats have had discrepancies with business interests; for example, they imposed new taxes on financial operations in order to extract more revenue from the banking sector, which had benefited from market reforms. Similarly, the government imposed regulations on retirement funds administered by private firms. And more recently, an originally leftist-oriented government elected in 2011 has maintained technical continuity in economic institutions and is using experts' knowledge to extract more resources from mining enterprises. So, despite being closer to business than are their Colombian counterparts, Peruvian experts still remain quite autonomous from social forces and will likely find themselves more at odds with the private sector when the current period of economic growth passes.

A second limitation of a business-agents perspective is that it cannot explain experts' considerable autonomy in social policy sectors such as health. My research documents that experts in these areas with weak or no links to economic elites effectively opposed political and IFI interests. The Peruvian case, for example, shows that when business actors promoted particular health reforms, such as privatizing workers' health insurance, experts in the Ministry of Health ignored them and adopted different policies. Similarly, the 1993 Colombian health reform included a series of benefits for informally employed Colombian citizens that increased the social obligations of private enterprises and thus clearly negatively affected business interests. More recently in Colombia, health technocrats are pushing for stronger regulations and control over health purchasers.

To conclude, although businesses muster considerable power in the region and have close linkages to experts, a "business-agents" perspective fails to explain the role played by experts in both countries. Significant discrepancies exist between experts and business interests, showing that technical policy makers have meaningful autonomy from economic elites. Even if this perspective is insufficient to explain technocratic power, however, my theory does recognize the strength of business actors as an important factor behind variation in technical autonomy across policy areas. Expertise is indeed more likely to become entrenched in sectors of great interest for business elites, such as economic areas, while it remains more vulnerable in social sectors with no similar stakeholders, such as health. Yet, while I argue in the next chapter that business stakeholders enhance technical autonomy and contribute to its entrenchment, a business-agent perspective does not fully explain technocratic power in the region.

CONCLUSION: THE POWER OF EXPERTS

In sum, the three instrumentalist theories do not capture the role played by experts in Latin American democracies. Their main limitation is that they all exaggerate the control of a particular actor over technocrats. *Técnicos*' position in the state allows them to frequently design and implement policies not always understood or promoted by politicians, IFIs, or businesses. As such, experts are better conceptualized as autonomous actors.

By underestimating technocrats' relevance, these instrumentalist perspectives miss one of the most important sources of policy reform and institutional change in contemporary Latin America. These instances cannot be dismissed as anomalies or exceptions within regular patterns; in recent decades, technocrats have regularly been responsible for selecting and implementing significant reform packages without significant input from other actors. More broadly, as discussed in the conclusion, these findings point to the limitations of electoral-incentives theories and structural approaches for the study of politics in Latin America: these theories do not give enough attention to powerful state actors such as technocrats. In the jargon of political science, technocrats deserve to be considered independent variables when studying politics and policy making.

The previous discussion also shows that technocratic power in some policy areas is more stable than is usually assumed by these competing theories. Some of these perspectives tie experts' influence to an acute crisis that weakens politicians and makes them delegate authority to experts. My findings confirm that expert influence rises under crisis conditions but, as discussed throughout the cases, is not dependent on these conditions; it remains stable even once the crisis has passed. In what follows, I show that economic technocrats in Colombia and Peru, and even health policy experts in the second Colombian episode, have remained quite influential under "normal" political and economic conditions. The Colombian technocracy, for example, advanced market reforms as a preemptive strategy – not in response to a crisis. Similarly, Colombian and Peruvian experts have maintained their preferred policy of fiscal prudence even under booming economic conditions, a time during which any autonomy conferred solely by a crisis situation would surely have dissipated.

These findings open a new set of questions that instrumentalist theories ignore or downplay. What is the source of technical autonomy? Why are powerful technocrats more likely to be found in the economic policy arena rather than in other policy areas where expertise is also required, such as health policy? And why does technical autonomy consolidate in some areas while remaining more precarious in others, vulnerable to changes in the political landscape? In the next chapter, I develop a theory of technocratic autonomy that seeks to answer these questions.

3

A Theory of Technocratic Autonomy

In the previous chapter, I showed that instrumentalist theories fail to account for the autonomy of technocrats in several Latin American states. Under certain conditions, *técnicos* can act as agenda setters who draft policies according to their own preferences and, often, successfully implement them, even against the desires of politicians, IFIs, and business interests. Furthermore, in some policy areas experts even achieve entrenchment – meaning incumbent politicians are informally forced to appoint officials in that policy area with similar levels of training and policy preferences.

As such, technocrats exercise equivalent powers to other political actors considered autonomous in political science, such as formally independent state agencies and international organizations (Krasner 1984, 224; Skocpol 1985, 9; Brint 1990, 369; Carpenter 2001, 15; Barnett and Finnemore 1999). As mentioned, these findings are surprising: technocrats usually lack either institutional protections, like bureaucratic tenure, and resources associated with political power in Latin America: votes, guns, or money. It seems that technocrats should be beholden to the politicians who appoint them or the societal and international actors who influence politicians' choices. Yet we observe experts enjoy considerable autonomy from these actors. How can technocrats be both autonomous and influential political actors in the region? In this chapter, I propose a theory of technocratic autonomy that explains (1) where experts' autonomy comes from, (2) why it varies across policy areas, and (3) what causes it to become entrenched in some situations and remain vulnerable in others.

The chapter is organized as follows. In the first section, I explain how expertise provides technocrats with considerable leverage over other sociopolitical actors. Then I account for the variation in technical autonomy across time and policy area by showing how two factors related to the politics of policy making in areas where experts operate favor technical autonomy. These two factors are crucial for explaining higher levels of technical autonomy, as they are especially important in limiting political intrusion, thus favoring technical continuity. Two additional factors related to the technical aspects of policy – the degree of technical consensus and the complexity of technical knowledge – further increase technical autonomy. In the third and last section, I show how these factors contribute to technical entrenchment and briefly discuss the conditions that may lead powerful experts to lose their autonomy. Throughout the chapter, I include evidence from episodes of technocratic autonomy in Colombia and Peru to substantiate my claims, although these points are further developed in the following chapters.

THE SOURCE OF EXPERTS' AUTONOMY: KNOWLEDGE IS POWER

Expertise is the fundamental source of technocratic autonomy. A scarce resource in Latin America, expertise allows technocrats to advance their policy preferences against the interests of other powerful actors. Diverse works in the social sciences highlight technical expertise as a source of political power. These works show how bureaucracies, state bodies, international organizations, and technocrats use expertise to gain considerable influence over other actors. There are two main reasons in the literature that explain why technical expertise is an effective political resource: (1) the growing legitimacy of technical knowledge in modern societies and (2) the leverage expertise provides to its user.

Some authors argue that the need to respond to complex social, political, and international demands in increasingly intricate societies gives technical actors strong legitimacy that translates into power (Meynaud 1969; Bell 1973; Haas 1992, 7–9). One piece of evidence for this growing social legitimacy is the increased importance of technical education for obtaining positions of power (Putnam 1977, 384–385). Technical knowledge provides legitimacy through the claim that its application produces objectively effective and politically neutral solutions to social problems (Schaar 1984, 120; Weber 1946). Because technical knowledge purports to follow scientific rules of internal coherence and thereby be impartial, technical actors may be less vulnerable to accusations of partisanship, even if values or ideological preferences underlie their assessments. Weber, for example, points out that bureaucrats, even those with hidden normative agendas, gain their authority partly from claiming they are rational and impartial actors who are best suited to solve complex issues (Weber 1946). In short, technical knowledge by itself provides considerable power to its user.

The increasing legitimacy of technical knowledge certainly contributes to experts' power in Latin America but does not alone fully explain their heightened influence. This legitimacy is relevant during moments of uncertainty, when politicians are forced to look for legitimized forms of expertise that can provide solutions to acute crises. However, my research suggests that technical legitimacy is more of an outcome than an intrinsic asset valued socially. Technocrats are not powerful just because they have social legitimacy, but rather gain this legitimacy over time due to the results of their policies. While this accumulated legitimacy further empowers technocrats as time passes by making it easier for them to discredit non-technical alternatives to their proposals, it does not fully explain experts' initial source of power and the crucial years in which their position is disputed.

A more relevant source of the political power of technical expertise is the leverage it provides over other actors. Because technical expertise is highly specialized and available only in those who invest significant resources in obtaining it, there is a strong information asymmetry that empowers those who possess it over those who do not. Thus technical knowledge not only grants experts legitimacy by enabling them to fulfill their duties with greater effectiveness than can other bureaucrats, but creates an information asymmetry that they can use to advance their own interests and influence. Knowledge provides strong power to its bearer.

One example of how technical knowledge provides influence comes from Weber, who argues that bureaucrats can exercise political power by supplying politicians with only selective information that leads politicians to adopt the bureaucrats' preferences (Weber 1946). Similarly, principal-agent theories in American politics focus on this asymmetry of information and uncertainty about policy choices to explain how bureaucrats can escape the political control of their formal principals (Miller and Moe 1983; Moe 1984; Waterman and Meier 1998).[1] Barnett and Finnemore highlight the way in which international organizations exploit their expertise to influence states. By classifying information according to their own preferences, adopting particular interpretations of international rules, and promoting the diffusion of certain norms and ideas, these organizations affect the interests of international actors and enhance their own influence (Barnett and Finnemore 1999). Lastly, Meynaud documents how high ranking bureaucrats in France influence their political superiors through diverse strategies, such as delaying regulation, using information for their own advantage, or implementing legislation selectively (Meynaud 1969, 74–87).

This asymmetry of expertise between experts and politicians is considerable in Latin America. In much of the region, politicians lack the technical capabilities and institutionalized bureaucratic support of their counterparts in developed countries (Santiso and Whitehead 2006, 19–20). Whereas politicians in the United States and Europe may have access to experts employed in their party or elsewhere in the bureaucracy who can evaluate technocrats' proposals and actions, the relative scarcity of such expertise in Latin America makes politicians more dependent on experts. Therefore, these stark technical asymmetries endow technocrats with considerable autonomy, even if they lack

[1] Most of this literature does not recognize bureaucratic autonomy as conceptualized in the other examples. These authors explain bureaucratic "independence" by rational, calculated, political delegation, thus limiting the claim that bureaucrats can have real autonomy.

the institutional protections enjoyed by high-level bureaucrats in developed democracies.

These information asymmetries explain why technocrats reach their positions in the state in the first place. Electoral-incentives perspectives correctly identify the most frequent mechanisms behind experts' appointment within the state: politicians need technocrats to respond to challenges they cannot handle by themselves or conduct complex reforms that will reap them political gains. The challenges that prompt the experts' appointment include economic crises, external demands from IFIs for reforms that require technically qualified personnel, and urgent domestic political demands.

The incumbent frequently appoints experts in order to signal to voters and other political and economic actors that the government is handling the challenge responsibly. As Haas discusses, "Failed policies, crises, and unanticipated events that call into question their [decision makers'] understanding of an issue area are likely to precipitate searches for new information, as are the increasing complexity and technical nature of problems" (Haas 1992, 29). Politicians faced with political quandaries will turn to specialists "to ameliorate the uncertainties and help them understand the current issues and anticipate future trends" (Haas 1992, 13). Experts provide authoritative recommendations and ideas about the best way to deal with these issues, using instances of *técnicos* in other countries that successfully responded to similar events to support their positions. They also have the skills to respond to external demands for high-quality policy making and to become the interlocutors of IFIs.

But politicians do not seek experts only in times of crisis. As political-incentives theories also highlight, politicians can use experts as a tool for political gains. When politicians can benefit from policy reforms, they often create technical "pockets of efficiency" to implement these changes, insulating high-quality bureaucracies from socioeconomic pressures and providing them with resources normal bureaucracies usually lack (Geddes 1994; 1990). Reformist politicians may also seek specialists to advance their sponsored reforms. As is clear in the Colombian health policy episodes, health experts reached the state level thanks to the genuine reformist goals of Presidents Carlos Lleras (1966–1970) and Gaviria (1990–1994). Both presidents embraced reformist agendas that transcended political gains. Whether because of need, ambition, or true concern for their citizens' well-being, incumbents find in experts the necessary technical knowledge their political parties and state bureaucracies lack.

Once in their positions, however, technocratic reform teams' expertise allows them to escape the limits politicians initially place on their power. Technocrats have a set of goals that do not necessarily coincide with those of their political superiors. The strong asymmetry of technical knowledge between experts and politicians, in addition to the fact that many *políticos* lack a proper understanding of the broader and longer-term effects of many reform proposals, allows experts to outmaneuver their formal principals. I identify two mechanisms through which experts gradually gain leverage over their political superiors and expand their control over policy areas: (1) the rise in technical asymmetry as a result of the increasingly complex policy that experts design and (2) experts' intentional efforts to augment their autonomy.

The first mechanism highlights how the act of appointing *técnicos* makes politicians more dependent on their skills. This mechanism is a direct consequence of delegating policy making to experts: *técnicos'* capabilities immediately increase the technical understanding required to manage policy. As different experts interviewed for this book pointed out, just by using the technical expertise their political superiors and fellow bureaucrats lack, and consequently raising the complexity of policy making, *políticos* would have difficulty managing these policy areas were they to wrest control of them from the experts. (interviews with Caballero Argáez 2007, Iwasaki 2007, Meloni 2007, and Pasco Font 2007). The result of politicians' dependence on experts is greater technical autonomy.

Gradually, the policy areas where experts are in charge become sufficiently complex that politicians lose familiarity with their internal operation and are no longer capable of running them swiftly. Drafting budget allocation plans, fulfilling transparency requirements, implementing internationally funded projects, and following internationally established administrative procedures are duties frequently beyond politicians' abilities, thus enhancing experts' leverage over their superiors. Experts also give politicians access to wide networks of international technical institutions and agencies with which experts share a common language. Drafting proposals for external funding or responding to particular demands from these institutions are technical jobs that politicians frequently cannot handle by themselves.

Two aspects amplify this dependence. First is the relatively short term of most ministers in Latin America. Ministers usually understand that their tenure in office will not last long. Therefore, they have no time to waste if they want to make any difference so they ask experts for advice about the best way to achieve their goals. Similarly, politicians take care to avoid causing unintended damage to ongoing projects and reforms, especially if the reforms are sponsored by international organizations. As discussed below, the higher the potential political costs of interfering with experts' policies and the complexity of the reforms underway, the higher this political restraint. In short, ministers are transient actors with little incentive or ability to challenge technocrats working within their ministries.

Second, in countries where political parties are weak, and thus professional politicians scarce, this asymmetry is even greater. Amateur politicians lack both basic information about how the state works and support from institutionalized political parties that have state management experience. The result is that politicians are more uncertain about the effects of changing policy, which provides more room for experts' stability. So, by acting in line with expectations, experts entrench expertise within the state and make other actors, political or socioeconomic, accept the new technical standards in policy making.

A crucial point in my theory is that experts play an active role in securing their autonomy. The second mechanism reveals how, using their superior technical knowledge, experts often strategically maneuver to protect and increase their autonomy from their political superiors. Diverse authors have highlighted this strategic use of expertise to advance technocrats' preferences in Latin America. Guillermo O'Donnell, for example, shows how highly trained individuals, tired of politicians' profligate behavior, formed alliances in the sixties and early seventies with the military in the Southern Cone to impose "rational and efficient" development models (O'Donnell 1973, 79–89). These experts brought to this alliance their internationally legitimized developmental recipes and external networks. Focusing also on the Southern Cone, Biglaiser documents how economists used their technical knowledge to convince the ruling military leaders of the soundness of their economic prescriptions and exclude less technically capable actors from cabinet debates (Biglaiser 2002). Similarly, Grindle documents how politically savvy Mexican experts under the *Partido Revolucionario Institucional* (PRI) used their knowledge to carefully advance their preferences, seek alliances with politicians and other bureaucrats, and protect their office's independence (Grindle 1977, 416–419).

Building on this previous work, I identify and further develop four strategies that experts employ to maintain and strengthen their influence over policy making. These consist of (1) purposefully excluding actors who lack technical training from policy making; (2) diverting political and socioeconomic pressures from their policy areas; (3) building informal and formal institutions that guarantee technical stability and reproduction; and (4) establishing alliances with other actors to advance their own preferences.[2]

First, technocrats use their knowledge to exclude actors without similar technical expertise from participating in policy debates. As mentioned above, sometimes this outcome is simply the natural consequence of appointing experts, but in other cases, it can be the intentional result of experts using their knowledge as a political weapon. Increasing the level of technical knowledge required to understand public policy gives experts a crucial advantage over other state actors, such as politicians and bureaucrats, and even private actors that may try to influence policy making. Experts, for example, present their arguments with studies that use empirical data to evaluate the consequences of policy choices. If politicians and other actors such as business groups cannot produce proposals of a similar technical sophistication, they are at a disadvantage in deliberations.

Second, expertise becomes a key resource for diverting political or socioeconomic pressures that threaten to override technocrats' preferences. Instead

[2] I focus here on strategies to maintain and enhance technical autonomy, but other strategies are also used by technical reform teams to advance policy reforms in developing states (see, for example, Grindle 2000; 2004). For more about reformist strategies to minimize political costs in the developed world, see Pierson 1994.

of clashing directly with their political superiors or socioeconomic actors by rejecting policies these actors do not support, experts use more subtle, yet highly effective, diversion strategies (Centeno 1993, 322–323; Grindle 1977, 421–422). My research identifies several of these strategies. Specifically, experts' control over information allows them to restrain or promote certain courses of political action. For example, experts may convince politicians or IFIs that particular electorally driven policies are infeasible or counterproductive and then commit their politically appointed superior, whose time horizon tends to be short, to long-term projects that have significance only *técnicos* themselves fully grasp. Similarly, experts can adapt laws to their own preferences by interpreting their scope and intention in ways that either narrow or widen their intended effects. Experts can also choose to selectively enforce laws in ways that harmonize these laws with the *técnicos'* original preferences. These strategies maintain and advance technical autonomy.

Third, experts often try to establish formal and informal institutional requirements to guarantee technocrats' continued appointment and reproduction. By demanding technical qualifications in order to gain office, they turn expertise into a requirement for recruitment into the agencies under technocratic control. By doing so, experts force politicians to hire others with similar levels of education and preference while preventing their political superiors from using positions as patronage spoils. Additionally, by providing adequate incentives, experts can guarantee their reproduction in the state. Colombian economic agencies, for example, offer scholarships to their members to pursue graduate studies abroad. Technocrats purposefully designed these incentives to reduce politicization. Colombian economic experts even created civil society institutions, such as the think tank Fedesarrollo, to enhance the level of expertise in society, provide support for expert-sponsored policies, and harbor experts when they leave the state (Botero 2005). Experts also provide formal institutional protections to agencies created to carry out their sponsored reforms. In the sixties, Colombian economic experts insulated export-promotion agencies (Juárez 1993a; Díaz-Alejandro, 1973). Similarly, experts in Peru during the 1990s created new regulatory agencies with management laws akin to those of private institutions, in order to prevent their bureaucratization (Wise 2003).

Finally, experts build strategic alliances with politicians, socioeconomic actors, and IFIs. As my research shows, technocrats may approach economic and political elites to ask for support for their policy preferences and reforms, sometimes enlisting these actors as allies against their political superiors that oppose these changes. Technocrats may even use the media, sometimes off the record and without their political superiors' permission, to create negative public perceptions of policies they oppose. Experts' knowledge is a critical contribution to these alliances, as other actors frequently lack the expertise to defend these proposals on technical grounds.

Expertise, then, is the main source of technocrats' autonomy. These findings are noteworthy and diverge from both structural and political-agents' theories that downplay the importance of expertise as an independent factor in policy making and politics. Nonetheless, I also find that the degree of autonomy varies across policy sectors and even within the same policy sector across time. In some policy areas, expertise is more effective as a political resource, allowing experts to have considerably more capacity to prevent external intrusion in their policy areas and advance their sponsored policies. Yet in other areas, technical knowledge seems to provide experts little protection from political interference. Even if technocrats are actors in all these areas, they are considerably more constrained in some than others.

Furthermore, in policy areas with favorable conditions, experts are more likely to achieve technocratic entrenchment, which occurs when the political cost of appointing unqualified individuals to head these policy areas becomes too high for the incumbent. Incumbents may appoint new officials, but they must also be experts who have similar technical credentials as their predecessors. At that point, only a major departure from "politics as usual" could lead to experts' removal from their positions. In other policy areas, however, this autonomy remains weaker, experts more vulnerable, and entrenchment unlikely. The passage of time alone does not necessarily make technical autonomy more secure in these areas. What factors explain this variation in technical autonomy across policy areas and in the same policy area across time?

DETERMINANTS OF TECHNICAL AUTONOMY

I propose that four factors – two related to the politics of policy and two related to the technical aspects of policy – determine the degree of technical autonomy experts will achieve in a given policy sector and the likelihood of technical entrenchment. The first two factors explain why some areas have higher technical autonomy and why technical entrenchment is more likely. The latter two, if present, enhance technical autonomy and raise the likelihood of technical continuity. The first two highlight the importance of politics and socioeconomic actors for technical autonomy and technical stability; the latter two support the argument that knowledge itself is a valuable political resource.

The Politics of Policy

Experts' knowledge and strategies account for a significant portion of their power in Latin America. Nonetheless, the political context in which technocrats operate affects the degree of autonomy experts secure and, as discussed in the next section, the likelihood of entrenching technical autonomy. Two aspects of the surrounding political situation are particularly important: (1) the cost to the incumbent of poor policy performance and (2) the constellation of stakeholders with interests in a policy sector. When poor policy performance has high political costs, such as when there are urgent problems to be solved, and there are multiple powerful socioeconomic actors competing to influence policy, all players in the political game may find that their optimal strategy is to not interfere with technocrats implementing "efficient" policy, which in turn gives experts space to establish their autonomy. In short, these politics of policy factors allow us to identify more likely and less likely cases for technical autonomy.

Political Cost of Poor Policy Performance

The first factor is the cost to the incumbent of delivering poor policy outcomes. As mentioned before, politicians usually appoint experts to respond to acute crises they cannot solve by themselves. After this initial appointment, incumbents evaluate the costs and benefits of leaving experts in control of policy making. I propose that, when potentially high, these costs decrease experts' vulnerability and enhance their chances of securing long-term control over policy making. Incumbents may push experts to adopt policy preferences that politicians favor, but they have reasons not to significantly reduce technical capacity or interfere in technical policy making. If politicians appoint officials explicitly as technocrats, it is paramount that these officials are seen as technically credentialed and independent of political control. Otherwise, politicians will face significant political costs. Thus incumbents refrain from interfering in these areas, enhancing autonomy and favoring technical continuity.

I argue that the most important political costs of undermining or firing technocrats are the potential consequences of hurting the population. Inadequate policy outputs in some issue areas can hurt the electoral chances of politicians, especially if poor policy exacerbates a crisis. If negative consequences are difficult to foresee, politicians are even more risk averse. And if a previous crisis has made the population more sensitive about policy making in particular issue areas, the potential costs of interfering are considerably higher. In other policy spheres, by contrast, inadequate policy outputs have less serious repercussions for politicians, and changes in personnel do not carry similar risks.

Regarding the political benefits of controlling a policy sector for the incumbent, the most important is the potential patronage resources it offers. Technocratic management reduces political discretion over the allocation of public resources. Some issue areas offer vast amounts of resources to political actors: control over budget distribution, large numbers of jobs to allocate among party members, well-paid positions in state-owned enterprises for political allies, procurement contracts for potential donors of campaign contributions, etc. Sectors with more resources to distribute, then, are tempting targets for interference by politicians.

When a new incumbent reaches power, the sitting incumbent that appointed experts changes his preferences, or when technocrats' preferences clash with those of politicians, experts' continuity is questioned. If there are no clearly perceived costs of intruding in a policy area, the incumbents' incentives to

TABLE 3.1. *Political Costs of Poor Policy Performance*

High Political Costs	Low Political Costs
• Considerable share of the population affected. • Mobilized/influential social classes affected. • High potential damage. • Visible damage that is easily traced to the incumbent. • **Prediction:** Incumbents refrain from interfering in these areas, enhancing autonomy and favoring technical continuity.	• Small share of the population affected. • Mainly lower, less politically influential social classes affected. • Low/diffuse damage. • Difficult to link the incumbent with negative policy outcomes. • **Prediction:** Few political costs for incumbents if they interfere with technical policies.

intervene and fire experts are higher. What factors are relevant for incumbents' evaluation of their political costs?

As shown in Table 3.1 four aspects seem especially relevant for this evaluation. The first is the share of the population affected by poor policy decisions. In general, politicians will find political decisions that affect a larger number of the population more costly. Secondly, if the policies in question hurt the upper and middle classes, which are typically more politically mobilized and influential than are poor sectors, potential political costs increase considerably. Third, the severity of the potential damage matters for this assessment. Poor decisions that could inflict serious damage are more costly than those with less severe impact. Finally, politicians will consider how visible and traceable the potential consequences are. In some issue areas, poor decisions, or stalling in lieu of making a decision, can have immediate and noticeable effects on the well-being of the population. If failed policy decisions have an immediate impact, political interference in technical policy making becomes more costly. However, if negative consequences are likely to appear only after a significant delay or if inaction is not politically noticeable, the incumbent has more freedom to interfere in policy making and override technocrats.

Given the factors that affect the political consequences of poor policy are relatively constant, the cost of intervention in an issue area remains fairly similar across time. However, certain conditions can vary over time, thereby changing the cost of firing experts. One such factor is the occurrence of crises, which can affect the magnitude of political costs in several ways. Crises can make the population more sensitive to inadequate policy making. The memory of crises that brought on pain and resulted in enormous costs to the population makes citizens more aware of the importance of some policy sectors, thus increasing the political costs of intervention. If experts are considered responsible for restoring social and economic stability, they would be associated with these better policy outcomes and thus more costly for politicians to remove from the state level. Similarly, if experts are blamed for the crisis, the political costs of firing them are lower.

Experts' sponsored policies can also raise the political costs of intervention in technical areas by producing positive feedback, defined as "the ways in which previous policy choices influence present political processes" (Pierson 1994, 39). As highlighted by Pierson, policies are not only an outcome of political struggle and negotiation. Policies can affect politics: "traditionally, researchers treated policy as the result of political forces (the dependent variable), but rarely as the cause of those forces (the independent variable). In the past decade or so, this has ceased to be true. Scholars working on a range of empirical issues have begun to emphasize that 'policies produce politics'" (Pierson 1994, 39). Unions, for example, can grow in number and become more politically salient due to regulations establishing mandatory registering. Similarly, social policies can group together recipients that will act in defense of their newly acquired benefits.

If the effects of experts' sponsored policies bring large portions of the population into contact with the state, the perceived political costs of poor outputs increases. If policies now affect a large number of vocal citizens (such as middle and high classes) who cannot "exit" these new regulations, then state action, or inaction, becomes more relevant for the citizenry. As such, technical autonomy is less likely to be violated. Politicians feel obliged to signal the population their commitment to responsible policy by appointing technocrats or keeping experts in their positions.

Important differences between economic and health policy exist with respect to these political costs. Mistakes in economic policy carry heavy and immediate political costs (Kaufman and Nelson 2004, 13).Typically, when the economy is well it doesn't garner attention. But the moment an economic crisis begins to brew, it catches the attention of the population. The public in general suffers the consequences of economic irresponsibility, a lesson widely learned in the region after the hyperinflationary crises of the eighties. Rich and middle classes cannot insulate themselves from the consequences of these poor policies. Furthermore, as discussed later, influential societal actors, such as the private sector, that have a major stake in economic policy will raise their voice privately and in the media if the economy is not properly managed. The political opposition can also exploit this political outcry. Therefore, politicians think twice before appointing political cronies or under-qualified individuals in technical areas. Even if using economic ministries for patronage remains attractive to politicians, the hyperinflationary crisis of the late eighties raised the perceived costs and risks of politicizing management of the economy. In Peru, the catastrophic effects of hyperinflation taught Peruvians about the need for responsible macroeconomic policy. Incumbents prefer to claim credit for adequate management of a policy area than to risk their stability for potential and uncertain electoral goals. Due to widespread learning from bitter experience, contemporary economic technocrats faced fewer threats from the incumbent.

In contrast, public health remains a sector in which political intervention carries lower costs. Although public health affects many individuals in society, the poor have the most at stake (Kaufman and Nelson 2004, 13; Nelson 2004, 39–41). In the region, the wealthy are covered by private health systems and do not have to deal with public health, and those formally employed within the middle classes usually have some sort of insurance provided by state social security (O'Donnell 1996, 10–12). In Hirschman's words, rich and middle class sectors can "exit" to private services rather than demanding a better health service (Hirschman 1970). The costs of maintaining the status quo are also low, as citizens have grown accustomed to a poor-quality health system and do not blame the incumbent for this outcome. Poor health policy is difficult to trace to the incumbent. Further, epidemics are likely to affect more vulnerable sectors of the population, limiting the effect of health crisis over the perceived political costs. A health system in crisis has been an everyday fact of life for many Latin Americans, but rarely has it generated significant backlash for politicians. Therefore, the costs to the incumbent of delivering mediocre health care are relatively low.

Furthermore, the political benefits incumbents can gain from health reforms are uncertain because the positive effects of the reforms may take a long time to become apparent, while the costs of initiating reforms, such as antagonizing unionized health care providers, manifest themselves in the short term (Nelson 1999, 16–22; Kaufman and Nelson 2004, 11–14). At the same time, the incumbent can gain significant benefits from maintaining direct political control of the health system given its abundant potential patronage resources consisting of jobs for party loyalists and the politically targeted distribution of health services (Kaufman and Nelson 2004; Nelson 2004, 38–39). As a result, health is an area in which experts remain more threatened by politicians.

Nonetheless, policy feedback can contribute to enhance political costs in health policy sectors. Variation in this factor helps to explain the intermediate outcome in the second Colombian health policy episode, a less likely area for technical autonomy. By incorporating both the formally employed and the poor into a single health system, the Colombian health reform increased the political cost of delivering poor-quality health policy as formal workers were now concerned with public health and were unwilling to tolerate patronage or irresponsible politics in the health sector. Although this cost is not as high as in economic areas, these higher costs contributed to the significant continuity of technical expertise in the second Colombian episode.

It is easier to provide evidence that growing policy-area concerns within the population lead to technical appointments. Media attention as well as the saliency of the issue in polls can be linked with technical appointments, as seen in the second Colombian health episode when the area became more technocratic in the face of a growing financial crisis in the sector. But providing evidence that politicians restrained themselves from intervening in policy areas is difficult because the observable implication of such conduct is inaction.

By itself, inaction does not support such a mechanism. It could also indicate commitment to technical policy as part of responsible leadership, which is how most politicians in Colombia and Peru portray the decision to appoint experts and accept technocrats' proposals. When interviewed for this book, for example, former President Gaviria concluded that making sensible economic policy through experts was part of Colombian leaders' political culture (interview with C. Gaviria 2013). Although some leaders do have a genuine commitment to the common good, my point is that there is not consistently similar restraint across other policy areas. Something more than just politicians' "better angels" guides them when handling the economy, especially in the years when technocracy was not as entrenched as it is today in Colombia or Peru.

So we are left with the question: What is the evidence that leaders restrained themselves due to a risk assessment? Apart from several politicians who have accepted that concerns about the political consequences of poor policy do drive technical picks and respect for technical decisions, in the following chapters I also highlight how technical advisors on the campaign trail and ministerial appointments once in office are often portrayed by politicians as signals to the population of policy responsibility. Additionally, a change in conduct of leaders who were former critics of technical policies, but once in office supported technical continuity, also counts as evidence of this risk assessment. In the following chapters we find leaders that bowed to technocracy even if they were harsh critics of technocrats before they had won office.

In conclusion, experts are more insulated from political pressures in areas where public policy can cause immediate and noticeable effects that hurt large segments of the population. In those areas, experts can exercise greater autonomy that is less vulnerable to being violated following changes of government or clashes between experts and political interests. Although political costs tend to be quite stable across time, they also can change because of certain traumatic events such as crises, and policy itself can also raise the potential costs of intervention.

Constellation of Stakeholders

A second contextual factor that explains variation in the existence and degree of technocratic autonomy across state agencies is the constellation of stakeholders related to each policy sector. The balance of power among important actors plays a crucial role for enhancing technical autonomy and promoting its entrenchment over time. Particularly, a balanced constellation opens space for experts, allowing them to exercise their power with lower risk of retaliation.

I define stakeholders more loosely than do most theories of public policy. The conventional view is that stakeholders are citizens, interest groups, or associations benefited or harmed by particular policies (Graham et al. 1999, 9). I use "stakeholder" more broadly to refer to social, economic, or political actors with strong preferences and interests ("stakes") in a policy area. Under this definition, private firms and business associations are among the additional socioeconomic actors included as stakeholders. Similarly, IFIs that participate actively in the issue area and make policy prescriptions will also be relevant actors. Even the political opposition counts as a stakeholder due to its interest in preventing incumbents from politicizing policy sectors.

I argue that a plural and balanced constellation of powerful stakeholders enhances technical autonomy by (1) providing experts with more space to adopt their own policy decisions and (2) increasing the likelihood that experts' strategies to advance their policy preferences will succeed. As discussed in the following section, this balanced constellation of stakeholders also raises the chances that technical autonomy becomes entrenched over time.

Cross-pressures from a plurality of powerful stakeholders bolster experts' autonomy. The logic behind this argument is that as the fragmentation of powerful actors surrounding a policy sector increases, it becomes more difficult for a single actor to control experts, thereby opening space for autonomy. Diverse powerful stakeholders provide different and contrasting preferences regarding policy options, which balance each other out. Stakeholders act as veto players who prevent competing actors from controlling the policy area. In areas where this constellation of powerful actors is balanced, the incumbents are aware that pushing experts too hard to adopt their preferences may cause a backlash from influential stakeholders, such as the political opposition, business, or IFIs, who would ideally prefer to control these policy agencies themselves. Yet when such direct control is not possible, these actors see experts as a second best option, especially if the most likely alternative is political control by the incumbent. The informal but firm control these stakeholders exercise over the incumbent and each other provides insulation for technocrats. *Técnicos* benefit from this space and adopt decisions according to their own preferences with less fear of being punished by other actors.

Similar theories are used in political science to explain diverse instances in which political actors gain autonomy from other socioeconomic and political forces. Riesman argues that the existence of plural interest groups in the modern United States precludes the possibility that a particular group controls the government, and gives it considerable autonomy (Riesman 1950, 235–255). Rustow explains the adoption of democratic institutions in modern Europe as a consequence of the stalemate between aristocratic and labor forces. A democratic regime does not guarantee the preferred options of either labor or aristocratic forces, but it is a reasonable "second best" option for both sides (Rustow 1970). Similarly, pluralist theories of judicial autonomy find a balanced and competitive party system to be a necessary condition for the emergence of independent judiciaries that make decisions free of political influence in the developing world (Ginsburg 2003; Magaloni 2003; Bill Chávez 2004; Ríos Figueroa 2007). These authors stress that political competitiveness provides space for the growth and expansion of judicial autonomy as equally powerful political forces check each other, preventing any one from gaining control of the courts. More closely related to the topic at hand, Evelyne Huber argues that a balanced civil society, with "divisions in the dominant classes and pressures from subordinate classes," contributes to state autonomy in Latin America (Huber 1995, 173).[3]

A balanced constellation of stakeholders provides experts with enough political space to adopt their preferences. In contrast, where there is only one powerful actor, usually the incumbent, this actor has an easier time opposing technical policies that affect its interest. If defiance by experts is strong, the incumbent can punish these technocrats and fire them more easily. Similarly, an unbalanced constellation of stakeholders where powerful actors join forces to attack experts will also lower technical autonomy.

The first set of relevant stakeholders is the political opposition. My explanation is similar to political-incentives theories that highlight the relevance of plural and similarly strong political actors for state agencies' autonomy. These theories tend to correlate state autonomy and state capacity with the degree of competitiveness of political parties. From this perspective, a more competitive party system will allow state institutions to gain autonomy and become more professional by limiting political intervention in bureaucratic agencies. As mentioned, this literature argues that parity among political parties leads to autonomous judicial bodies. Similarly, Geddes argues that political pluralism contributes to professional and independent bureaucracies by reducing the importance of patronage in electoral politics (Geddes 1994). For Geddes, balanced political parties were responsible for the administrative reforms to professionalize the bureaucracy undertaken in Colombia, Uruguay, and Venezuela.

As we will see, the political opposition certainly constitutes a check on incumbents who attempt to clash with technocrats. Experts become a guarantee for political stakeholders that certain policy areas will not be used to the advantage of the incumbent. In Colombia during the National Front, or during the last decade in Peru, the political opposition saw economic experts as protection against a politicized economy. Nonetheless, the episodes I explore also show that political parity alone cannot explain technical autonomy within the state. Political pluralism may open more space for bureaucratic professionalism in some policy areas but other area remains prone to political intervention. As discussed in the Colombian episodes, institutional autonomy largely transcends the competitiveness of the party system: a bipartisan system as it existed in Colombia for decades did not lead to more state professionalism in the overwhelming majority of bureaucratic agencies.[4] In areas that revolve around the allocation of distributable resources, political competitiveness by

[3] Similarly, Michael Mann finds the source of the state's autonomy and power in the need to balance the diverse interests of social forces within a given territory (Mann 1984).

[4] Similar negative outcomes can be found in two of Geddes's cases. Neither Venezuela nor Uruguay provides enough evidence of bureaucratic independence and professionalism coming into existence across the state out of political party parity.

itself does not seem a sufficient condition to explain technical and professional state institutions. It appears that equally strong political parties may sometimes share patronage rather than professionalize the state. In sum, political forces are only part of the story.

Socioeconomic interests and international actors sharing interests in a policy sector also help explain technical autonomy. The presence of powerful economic actors with veto power over politicians limits the possibility of politicians using (or sharing) the spoils of policy areas. Two types of economic stakeholders are particularly important for explaining technical autonomy: business interests and IFIs.[5] As discussed in the previous chapter, structural perspectives are wrong to attribute to business interests or IFIs the power to control técnicos. But these actors do have policy preferences they try to advance, frequently in opposition to the incumbent. By doing so, they interact with the other stakeholders, balance out each others' preferences, and thus affect technical autonomy by limiting political intrusion. Experts become "second best" options for these actors. Technocrats are accountable to more "audiences." On one hand they have to take these plural interests into account, but at the same time. they also gain autonomy from these interests. Although experts may not implement the exact preferences of these actors, they are considerably better than allowing these areas to be politically controlled. Other issue areas lacking a balanced constellation of stakeholders are likely to remain unprofessional. Table 3.2 presents these differences.

The balance of power among stakeholders helps to explain the variation in technical autonomy found in my case studies. Since stakeholders such as IFIs and business sectors closely scrutinize economic policy making and limit incumbents' intrusion, experts in this area have more space to adopt autonomous decisions. Stakeholders lobby privately or voice their disagreement through the media when individuals do not fulfill technical characteristics.

Health, by contrast, is a sector where there are no similarly powerful stakeholders. Few actors demand that politicians adopt effective health policies or restrict them from interfering with technical policies. Apart from businesses directly involved with the health sector (often more interested in regulations and distributive goods than in general and adequate policy outcomes), business actors in general show little interest or involvement in health reform (Nelson 2004, 40–41). Without this pressure, politicians who clash with experts have little reason not to fire experts and pursue new policies more in accordance with their own interests.

The importance of stakeholders for the continuity of technical autonomy becomes clear in moments when experts are more vulnerable, such as during

[5] IFIs have somehow different "stakes" in a policy sector than do the other actors mentioned. Policy outcomes usually will not affect IFIs directly. Nonetheless, I still consider IFIs stakeholders (in my extended definition) as these institutions have a strong interest in the advancement of their preferred policies in line with their institutional goals.

TABLE 3.2. *Constellation of Stakeholders*

Plural and Balanced	Plural and Unbalanced	Hegemonic Actor
• Diverse powerful actors interested in the policy area. Stakeholders act as checks and balances for one another, protecting technical autonomy as "second best" option. • Experts can use stakeholders as allies in order to advance technocrats' preferred policies. • **Prediction:** Experts have more space to adopt their own policy decisions. Increases the likelihood of success of experts' strategies to advance their policy preferences.	• Some of the stakeholders have similar preferences and these preferences oppose those of experts. • Few stakeholders support experts' strategies. • **Prediction:** Technocrats are more vulnerable to intrusion.	• Experts are only accountable to one powerful actor, usually the incumbent. In the event of conflict, experts lack external support to safeguard their autonomy. • Few or no stakeholders to serve as allies of experts. • **Prediction:** Highly vulnerable to intrusion and dependent on the incumbent's will.

changes of administration or clashes with incumbents. The pressure on the incumbent to maintain technical management in economic institutions is evident during the days preceding the appointment of a new economics minister. As shown in the two economic episodes with high technical autonomy (Colombia 1958–2013 and Peru 1990–2013), stakeholders make it clear that they prefer a well-educated appointee with economic credentials who can guarantee technical quality and continuity. Particularly, when a technical minister is dismissed, economic analysts and risk assessment agencies immediately comment in the media about the dangers for stability if the incumbent appoints a minister without such credentials. Experts are guarantors of stakeholders' interests and these strong actors demand clear signs of continuity. Once a technical minister is nominated, these actors produce reports emphasizing how the pick is adequate for economic stability.

By contrast, the fall of health experts due to clashes with politicians usually goes largely unnoticed. In the absence of powerful stakeholders, these events are criticized mainly by a few other health experts who share the perspective of the ousted official. Other powerful actors and the public at large remain unmoved by these events. Therefore, the costs of firing experts and not replacing them with qualified individuals are much lower. In consequence, politicians' latitude to change policy and personnel in health agencies is higher than in economic ones. Clashes with politicians can result in a considerable loss of technical expertise.

The health policy episodes provide additional evidence of stakeholders' relevance for technical autonomy. The enhancement of health expertise documented in my the episodes in the 1990s was linked to moments in which IFIs and other international stakeholders promoted health policy reforms. By doing so they increased politicians' need for health expertise and allowed experts some autonomy for making decisions. As discussed in the Peruvian episode, IFIs' declining interest in health reform at the end of the nineties harmed experts' autonomy.

On the other hand, in the first Peruvian economic episode (1980–1983), I show how an unbalanced constellation of stakeholders can affect economic technocrats. Although experts in MEF initially had considerable influence, powerful political and business actors who opposed their orthodox economic reforms eventually defeated them. After the debt crisis and, especially, the economic meltdown of the late 1980s, a stronger consensus in economic policy and more balanced constellation of stakeholders emerged with the result of increasing the chances for technical autonomy. In clear contrast to the previous experience of the 1980s, technocratic autonomy in economic policy has persisted in Peru.

A second way in which a plural and balanced constellation of stakeholders increases technical autonomy is by making experts' strategies to advance their preferences considerably more effective. As mentioned, sometimes experts play stakeholders off against each other and establish alliances with them according to the issues under consideration. Experts can evaluate the preferences of various stakeholders and use them to counteract the pressure they receive from different quarters.

Several examples in this book show such strategic action. As discussed, when politicians are not responsive to their preferences, *técnicos* can secretly ask IFIs to exercise leverage over the government. Or they can ask these institutions to demand that only highly qualified individuals are appointed to ongoing reform projects, thus limiting political manipulation of externally funded projects. Experts can also establish links with business associations or even the political opposition to advance their proposals. Alternatively, they can seek the support of other social actors, such as the media, think tanks, or interest groups to counterbalance political pressures.

The range of potential allies is greater for some technocrats than for others. Economic *técnicos* have much more room to maneuver. Some business associations can ally with experts in the effort to overcome resistance to particular reforms emanating from politicians or other business sectors. For example, exporters can support experts' efforts to open the economy, against the interests of protectionist business sectors. Or IFIs can finance pilot projects, such as developing more transparent budgets or new systems of financial accountability, that the incumbent is not willing or interested in funding. In health, by contrast, the incumbent is the dominant interest. Usually only IFIs, international organizations involved in health policy, and some local NGOs have a sustained interest in the public health system. The result is that health experts have fewer stakeholders to play against one another.

Summing up, the previous two "politics of policy" factors explain variation in technocratic autonomy. These factors allow us to identify policy areas where technical autonomy is likely to be high and others where it is expected to remain lower. Elevated political costs and powerful actors with interests in a policy area enhance technocratic autonomy by limiting political intrusion. In contrast, if these political costs are low and the constellation of stakeholders surrounding a policy area is unbalanced or dominated by a single political actor, technocratic autonomy will remain more vulnerable. As discussed later, these two factors are not only fundamental in explaining technical autonomy but also entrenchment. In areas where these factors are absent, political clashes, or mere political rotation, can abruptly change existing patterns of autonomy. These two factors account for the broad patterns observed in the two policy areas I contrast in this work: economic policy and health policy.

Technical Aspects of Policy

Two other factors related to the technical aspects of policy also affect expert autonomy: technical complexity and experts' consensus. These factors augment experts' autonomy by limiting political intrusion in technical policy areas and enhancing experts' legitimacy. When these factors are absent, they contribute to weaken technical autonomy even in "more likely" policy areas such as economics. And where present, they contribute to autonomy in less likely policy areas such as health. Additionally, these factors contribute to explain variation in technical autonomy in the same policy area across time. After discussing these two factors, I explain why they are less crucial for technical autonomy than are the previous two.

Technical Complexity

Technical complexity also shapes variation in technical autonomy. Experts are more autonomous in sectors in which the level of technical qualifications required to understand policy making is higher. Complexity increases the asymmetry between *técnicos* and other actors, which is at the base of experts' autonomy, and thus limits intervention in these policy areas. Complexity also makes it more difficult to foresee the consequences of policy choices. As discussed above, if policy consequences are difficult to foresee, political intervention becomes more risky than in policy areas in which these consequences are easier to grasp (Barnett and Finnemore 1999; Nelkin 1975, 36–37). Politicians and sociopolitical actors have more difficulty evaluating the likely costs and benefits of policy decisions and thus rely more on experts' assessments.

TABLE 3.3. *Technical Complexity*

High Complexity	Low Complexity
• Specialized scientific knowledge enhances technical asymmetry between experts and politicians. • Technical complexity makes political costs more difficult to evaluate by politicians. The consequences of decisions related to the provision of public goods are more difficult to grasp by non-trained individuals. • **Prediction:** Higher thresholds for political intrusion in technical policy areas and enhanced expert legitimacy.	• Non-specialized scientific knowledge is easier to understand by non-technically trained individuals, such as politicians. • Potential political costs are easier to evaluate. Politicians are able to more easily grasp the consequences of decisions related to the distribution of particular goods. • **Prediction:** Political intrusion is more likely.

What makes some types of knowledge and policy choices more complex than others? As shown in Table 3.3 two aspects are relevant to evaluate this complexity. First, the more specialized the scientific knowledge needed to understand the available policy options, the more complex the policy area. This level of complexity partly depends on how experts believe the issue area can be analyzed scientifically and how generalized this belief is among other actors. The more prone to modeling and quantification, the easier it is for technical practitioners to exclude other non-trained actors in a policy area through the use of specialized concepts and vocabulary. In order to be considered scientific, these "languages" must adhere to norms of internal coherence. This technical jargon limits who can participate in policy debates. Statistics, econometrics, and mathematics are all powerful mechanisms to exclude uninitiated individuals who cannot counter experts' assessments with similar evaluations (Grindle and Thomas 1991, 61).

Similarly, the consequences of policy choices for particular goods are less complex and therefore easier to foresee than are the consequences of decisions related to public goods. Specifically, when the technical debates are focused on budget allocations to particular services and programs with identifiable beneficiaries, actors such as politicians have a better grasp of the immediate consequences of those decisions for their interests. This assessment is more difficult in areas in which each decision requires a high degree of specialization and decisions themselves have broader effects not traceable to particular beneficiaries. These complex decisions determine the future availability of resources and public goods, issues difficult to foresee by an untrained individual. In the following chapters I evaluate the degree of technical complexity by analyzing the type of knowledge demanded in order to elaborate policy proposals in the issue area under consideration, and to participate in debates about the content and effects of policies.

This factor is important for understanding the different levels of autonomy prevailing in economic policy versus those in health policy. Politicians have a harder time assessing the implications of economic policy decisions. Economic policy is a complex, highly formalized discipline that requires mathematical and statistical skills. Determining adequate responses to economic crises or predicting the precise consequences of public spending on inflation are all policy decisions that largely escape the understanding of politicians. These limitations become more evident in moments of uncertainty, when wrong decisions risk deepening the economic problems of the country. Economic policy remains obscure and inaccessible for politicians.

In contrast, health policy is more accessible to political actors and associations of professionals working in the health sector (e.g., doctors and nurses). Even if health policy has become more complex in recent decades, especially after the emergence of the highly technical health economics approach in the 1990s, it remains less complex than is economic policy with effects that are more easily understood. Debates about health policy often involve decisions affecting the allocation of divisible patronage resources, a topic on which interested parties will certainly have opinions. As such, it is more difficult to exclude political actors from health policy debates.

A policy sector's technical complexity can vary across time, thereby enhancing or reducing experts' ability to marginalize other actors from policy making. The episodes I discuss clearly show this variation. In the 1960s, Colombian economic experts had the advantage of commanding expertise in the country when no other actors had similar capabilities. By raising the technical level required for understanding policy making, Colombian experts gained bargaining leverage vis-à-vis other economic "experts" (usually lawyers) linked to political parties and business interests (Cepeda Ulloa and Mitchell 1980). In the 1980s, Peruvian experts lacked a similar advantage because there were trained economists with alternative perspectives who could question their reform proposals on technical grounds. The health policy episodes also show how complex knowledge excludes other non-technical actors from policy making. As mentioned, the spread of the health economics approach in the 1990s made health policy significantly more complex, marginalizing other actors such as politicians and doctors.

Experts' Consensus

A second factor linked to experts' knowledge that boosts technocratic autonomy is consensual expertise. Expertise becomes a more effective resource for gaining independence and influence when experts agree on what constitutes adequate policy. When there is strong agreement among experts regarding how to properly diagnosis social problems and about which solutions look promising, an "epistemic community" exists: "a network of professionals with recognized expertise and competence in a particular domain and an authoritative claim to policy-relevant knowledge within that domain or issue area"

(Haas 1992, 3).[6] This consensus enhances experts' leverage by making the technical policy options they advocate look largely unchallenged by other experts. Technical legitimacy is firmly grounded in the informed opinion of other experts. Disagreements among technocrats are limited to smaller points that do not threaten the dominant paradigm.

When there is consensus among technocrats, politicians are more constrained to appoint only technically trained individuals to policy-making institutions. If politicians make decisions that deviate from this consensus, such as appointing an unqualified individual, these decisions can easily be condemned as irresponsible or politicized. Even if politicians decide to replace particular experts, they are forced to appoint new individuals with similar preferences. Disputes with politicians, then, can lead to changes of individual experts, but not to the appointment of less-qualified individuals or a broad change in policy prescriptions.

In contrast, when this consensus is lacking, divisions within technical ranks can be used by political and/or societal actors to weaken technical autonomy. These actors have a choice and can select *técnicos* aligned with their own preferences, rather than having to accept the preferences that experts consensually hold. In this way, politicians limit experts' claim to hold positions representing what is commonsensical in their professions, and turn technical debates into political ones, with politicians using "their" experts to advance their preferred alternatives (Nelkin 1975, 40–41, 54).

Geddes notes that "in most countries, political leaders can find economic experts who will advocate a wide range of policy prescriptions." Thus, "if one set of policies leads to opposition, the economists who proposed it can be replaced by others who will advocate more politically palatable ones" (Geddes 1995, 210–211). Geddes is right to point out that this divided camp of experts is normal in diverse countries. But the episodes I explore show that the availability of experts with diverse views is not as universal as assumed by the author. Experts' consensus can limit political choices for alternative experts who hold visions closer to politicians' interests.

How do we know in advance whether or not there is technical consensus in a policy area? Two aspects are relevant to answer this question (Table 3.4).

[6] The full definition of an epistemic community includes the following specification: "Although an epistemic community may consist of professionals from a variety of disciplines and backgrounds, they have (1) a shared set of normative and principled beliefs, which provide a value-based rationale for the social action of community members; (2) shared causal beliefs, which are derived from their analysis of practices leading or contributing to a central set of problems in their domain and which then serve as the basis of elucidating the multiple linkages between possible policy actions and desired outcomes; (3) shared notions of validity – that is, inter-subjective, internally defined criteria for weighing and validating knowledge in the domain of their expertise; and (4) a common policy enterprise – that is, a set of common practices associated with a set of problems to which their professional competence is directed, presumably out of the conviction that human welfare will be enhanced as a consequence" (Haas 1992, 3).

TABLE 3.4. *Technical Consensus*

High Consensus	Low Consensus
• Less debate about the adequacy of implemented policies. • Limits the availability of alternative experts with policy options that are closer to the incumbent's preferences. • **Prediction:** Stronger legitimacy of experts' recipes and increased likelihood of continuity.	• Alternative technical responses to the same issue are available. • Experts with different policy option perspectives that are closer to the incumbent's preferences are available for the incumbent to choose from. • **Prediction:** Experts' legitimacy is weaker; higher possibility of intervention.

Strong disagreements between the leading practitioners of the discipline about the central aspects of policymaking and policy priorities, such as where to allocate resources, indicate low levels of consensus. In contrast, well-established consensus about policy issues, including what constitutes responsible policies or where to direct resources, indicates a considerable degree of consensus. There may be disagreements about other aspects of policy making, but a solid set that provides common ground for experts assures a more unified front in an issue policy area.

Technical consensus is of upmost importance for explaining the variation found across countries in economic and health sectors and over time in the economic episodes in Peru. In the 1970s and 1980s, there were clearly divergent approaches to economic policy in Peru, and most of Latin America more generally, which gave politicians the option of choosing from among different economic recipes. As a result, the fall from power of orthodox experts in Peru during the early 1980s was facilitated greatly by disagreements with heterodox economists. Politicians and business associations used other experts to delegitimize technocrats' reforms.

Nowadays, broad agreement on what constitutes adequate economic policy prevails in the region, partly as a result of the damaging effects caused by the debt crisis of the 1980s (Montecinos, Markoff, and Álvarez-Rivadulla 2010, 15–17). The same orthodox and heterodox experts whose approaches to economic policy had little common ground two decades ago now share many assumptions, goals, and policy instruments. Although other economic issues remain disputed (e.g., privatizations, labor law, the use of subsidies for promoting industries, etc.), this consensus among experts highlights the need to maintain fiscal equilibrium, control patronage spending, and avoid inflationary pressures (Silva 1991, 405–406; Centeno and Silva 1998, 11–12). Politicians cannot appoint alternative experts who would be profligate.

Considerable technical consensus in economics has existed in Colombia since the time of the National Front. This common ground among Colombian experts partly explains the early emergence of technocracy in that country. For more than forty years, the Ministry of Finance has pursued a generally consistent set of economic policies even under turbulent political conditions. In Peru, such a consensus emerged after the hyperinflationary crisis of the late 1980s when the excesses of heterodox policies discredited it as an alternative to neoliberal economic policy. In both countries, politicians are now informally required to appoint officials that are committed to this consensus. A purely political appointment is not likely to be tolerated.

In contrast to economic policy, there is more limited agreement among experts in the health sector (Kaufman and Nelson 2004, 12–13). Public health policy has a long history of strong disagreements about what constitutes an adequate approach. Up until the 1960s, health policy prescriptions recommended the creation of a single social security system to provide health care in developing countries. Starting in the late 1960s, a new approach challenged this traditional prescription by questioning the regressive effects of alleged universal systems: these "universal" policies focused on the urban and formally employed population, leaving out the worst-off in society, such as those living in rural areas or those informally employed (Ewig 2010, ch. 2). Gradually, this new approach earned the support of the World Health Organization (WHO) and became known as "primary health care." From the primary health care perspective, it is more efficient to prevent illness by improving the living and nutritional conditions of the population than to cure it later or invest in expensive medical equipment. As a consequence, the WHO recommended substantial reforms promoting preventive policies, communal work, and decentralized allocation of technical and human resources to those sectors most in need (Cueto 2004,1868). Under the motto "Health for All by the Year 2000," the WHO's 1978 Alma Ata declaration promoted primary health care globally (Cueto 2004).

But this approach did not remain unchallenged. Other experts questioned primary health care for its broadness and for lacking specific, measurable goals. A more restricted version, selective primary health care, emerged in the eighties. This perspective also proposed focusing resources on the needy, but left aside some universalistic goals of the original primary health care model (Cueto 2004). Health economics emerged as a natural complement to this approach: only economic knowledge could properly assess which sectors needed resources the most and then determine where they should be allocated. When the World Bank launched its influential campaign for public health reform in the late eighties, it embraced a restricted version of selective primary health care, stressing the importance of health economics for policy implementation (Ewig 2010, 68–73). In its leading publications *Financing Health Services in Developing Countries: An Agenda of Reform* (1987) and *World Development Report: Investing in Health* (1993) the World Bank prescribed focusing on those sectors of the population that could not be covered by the market (Ewig 2010, 69–71).

All these approaches coexisted in Latin America during the nineties (Franco 1996; Arroyo 2000, 73–75, 161; Yglesias 1999). Frequently, health experts disagreed on which solution or policy to privilege. Choosing priorities, however, becomes a political decision because certain social groups will benefit more than will others depending on the chosen policy (Kaufman and Nelson 2004, 12–13; Nelson 2004, 31). This lack of agreement about priorities leads to harsh criticism and conflict among experts, opening the door for political intrusion and sometimes dramatic changes from administration to administration. Evidence of these disagreements is found in the different evaluations of policy reforms in both countries. Reconstructing health reforms through interviews sometimes felt like watching *Rashomon*,[7] as the successes highlighted by one group of experts were described as failures by experts with a different ideological or policy orientation. These discrepancies among experts have affected the stability of technical expertise in health policy areas.

Although these conflicting views still exist, health experts' positions have converged in the last decade. A significant level of agreement has emerged around a soft version of selective primary health care. This nascent consensus is less solid than is the existing consensus in economics, but it nevertheless reduces conflict among experts. Even sectors that demanded broad and universal systems of health coverage in the eighties now recognize the constraints imposed by resource limitations (Nelson 2004, 29–30). Health economics has become increasingly important in defining and solving the resource limitation challenges of public health. This consensus is now partially endorsed by institutions such as the World Health Organization (WHO) and the Pan American Health Organization (PAHO), and has provided some common ground to these technical debates. This change explains why the conditions for technical autonomy in health policy within both countries are now more favorable than they were in the past, although disagreements are still more prevalent than they are in economic policy.

These two factors enhance technical autonomy by further reducing the incumbent's chances of politicizing a technical policy sector. Higher levels of technical autonomy are expected in policy areas of technical complexity and where there is expert consensus. By contrast, where these factors are not present, technical autonomy is more vulnerable to political interference. As discussed in the episodes I explore, these two factors contribute to explain why expertise failed to entrench in the Ministry of Economics in Peru in the early eighties, even though it represented a more likely case for technical autonomy. Additionally, they help us to understand why Colombian and Peruvian health experts achieved more continuity in the nineties than was expected in a policy area in which experts were less likely to experience technical autonomy.

[7] In Akira Kurosawa's Rashomon different characters narrate the same incident. Each one of the characters provides a very different account of this same incident.

Why are these two factors, associated with the technical aspects of policy, of less importance for technical autonomy than are the previous two? The primary answer is that they have little effect on autonomy where there are no perceived political costs of poor policy performance, or if the policy area lacks a balanced constellation of stakeholders. To begin with, no matter how much complexity or consensus, politicians would not have the urgency to strengthen policy areas in charge of these issues. And if they do appoint experts, they would not be worried about the political consequences of intervening in these areas because the population are not that interested in these policy areas. Further, the technical and scientific communities where experts come from do not present a similar impediment to external interventions as does a balanced constellation of powerful stakeholders, no matter how much consensus exists within these communities. Scientific offices and/or special projects of a highly complex nature can be dismantled with little public awareness and with no strong actors opposing these decisions. These factors by themselves do not favor the emergence of strong and autonomous policy areas nor ensure their long-term stability, thereby limiting experts' latitude to make policy, in the case that they do emerge.

TECHNICAL ENTRENCHMENT AND THE FALL OF EXPERTS

In areas where these favorable conditions are present, experts are more likely to gain technical entrenchment. Technical entrenchment occurs when politicians are informally constrained to continually reappoint experts in certain policy areas. When the precedent of appointing technical ministers in the economic area has become the norm, as is currently the case in Colombia and Peru, naming an unqualified individual to a high-level economic position carries extreme political costs; politicians need to signal their commitment to technical continuity or face the consequences. Speculation about who will become minister of economics usually focuses around a small number of acceptable individuals for the position. In contrast, policy sectors where these conditions are not present, such as health, are much less likely to produce such entrenchment, and technical autonomy remains weak.

Interestingly, the passage of time by itself does not produce this entrenchment in health policy areas. Peruvian health experts achieved considerable continuity but they remained vulnerable. Why do some experts achieve technical entrenchment while others fail to do so? And how do these areas with strong technical autonomy lose their insulation and professional technical level?

The aforementioned factors contribute to the explanation of why technical expertise becomes entrenched. Under these positive conditions, experts' strategies to increase the formal and informal requirements for expertise in policy areas are more likely to succeed. Technocrats can use their authority to adopt rules and regulations that enhance their power. In addition, they gain legitimacy as their authority becomes less disputed. Over time, other state actors also have to learn how to communicate in technical terms, further legitimizing technocrats' knowledge. Although when present the four factors contribute to technical stability and facilitate experts' strategies to gain autonomy, the first two factors – political costs of poor policy performance and constellation of stakeholders – are especially relevant to understanding why experts' position in the state sometimes produces entrenchment.

First, if a significant portion of the population perceives experts as guarantors of prudent policy, the political costs of firing these experts is high. Politicians' capacity to subordinate experts in this situation is therefore limited. If crises are linked to previous non-technical, patronage-driven policy making, then politicians are under close scrutiny to send a clear message of responsibility by maintaining experts in their positions. And as mentioned before, experts' sponsored policies can also expand the section of the population that is interested in experts' continuity, thus making entrenchment more likely.

Second, a balanced constellation of powerful stakeholders is fundamental to technical entrenchment. Different authors discuss how and why powerful actors become active defenders of institutional stability and continuity (Pierson 1994; 2000; North 1990; 1995; Levitsky and Murillo 2009; M. Johnson, forthcoming). As highlighted by Levitsky and Murillo, "When institutional arrangements persist (and are enforced) over time, surviving repeated crises and changes of government, actors develop expectations of stability and consequently invest in skills, technologies, and organizations that are appropriate to those institutions" (2009, 123). Similarly, North concludes that "political institutions will be stable only if they are supported by organizations with an interest in their perpetuation" (North 1995, 25).

Powerful stakeholders, such as business interests and IFIs, see technocrats as guarantors of institutional arrangements. Technocrats are responsible for enforcing and guaranteeing the respect of institutions. Technocrats act as credible commitments for stakeholders that policies will be predictable and the status quo maintained. Technocratic stability makes stakeholders expect policy stability and act accordingly, thereby giving these actors greater stakes in institutional and organizational continuity. They become more vociferous defenders of technical continuity and the status quo, producing more support for the prevailing institutional arrangements (Pierson 2000, 259). As such, technical areas are also informally protected by these powerful stakeholders that are interested in institutional continuity.

As mentioned previously, experts' policies can enhance the power of some of these stakeholders, or even create new stakeholders, increasing the constellation of actors supporting technical autonomy. Policy can change the balance of power among different interest groups in society, enhancing the power of those who benefit from experts' regulations and reducing the influence of those negatively affected by it (Pierson 1994, 43–44; 2000, 259). These changes in political actors' power can contribute to technical continuity.

As discussed in the second Peruvian economic episode, the continuous application of market reforms has increased the power of business sectors privileged by this new institutional arrangement: exporters, mining interests, and financial institutions. This greater influence has produced stronger private actors, while politicians have simultaneously lost power in the country. As such, the support for experts has increased. Similarly, in the second health policy episode in Colombia, health reforms created new stakeholders (health providers and purchasers) within the system that counterbalanced the power of the incumbent and opened up more space for technical autonomy.

Nonetheless, even in cases where technical autonomy is stable, it can, under certain conditions, reduce drastically as technocrats succumb to political conflict. In Argentina, technocrats in the Ministry of Economics were considered strong and influential, only to be dismissed by Néstor Kirchner in the early 2000s, after a series of events that had weakened their position. Kirchner considerably subordinated the ministry to his political interests by appointing less qualified personnel and taking away some of its power, a trend followed by his wife and successor Cristina Fernández de Kirchner. Similarly, following his election in 2005, Evo Morales politicized the Bolivian Ministry of Economics, widely held as a technical policy area (although, as discussed in the conclusion, to a considerably lesser degree than in the Argentinean and Venezuelan cases). Finally, and although not an economic case, during the bipartisan *Pacto de Punto Fijo,* democratic Venezuela witnessed a highly technical bureaucracy emerge in PDVSA, the state oil company. These bureaucratic elite were dismissed in 2002 after Hugo Chávez clashed with the agency during his second government.

How did these technical actors lose their power even though their autonomy was widely seen as stable? In the conclusion, I discuss more thoroughly these three episodes of technocratic demise to illustrate the conditions under which even entrenched experts can lose control over policy. The two politics of policy factors are also fundamental in explaining these outcomes. First, an acute crisis traced to technocrats' policies can reduce the political costs of dismissing experts, allowing politicians to intervene in technical policy areas without having to face high political costs. If experts with alternative policy prescriptions are available, the decision to intervene is even easier. And if economic conditions ameliorate (e.g., a positive change as a result of external economic conditions), the risks of intervention for the incumbent are even lower. Second, when the constellation of stakeholders becomes unbalanced, either by a concentration of power by the incumbent or an alliance of several powerful actors that oppose experts' policies, experts become more vulnerable. If the incumbent concentrates power, and IFIs and business actors weaken or have fewer mechanisms of pressure on the government, experts are vulnerable. While each of these factors can have independent effects, they can also occur simultaneously. Under such conditions, we can witness abrupt technical decline.

These cases of technocratic demise show how acute social problems that delegitimize experts' policy prescriptions and a fundamental change in the constellation of stakeholders allow new incumbents to push aside existing experts. Experts fall from power under such negative conditions, and they are typically not replaced by actors with similar expertise or autonomy. Policy areas become more politically dependent.

CONCLUSION

The theory developed in this chapter seeks to explain experts' autonomy and power in democratic Latin America. The theory specifies the causes of technical autonomy as well as its variation across policy sectors and over time in the same policy sector. I discuss how expertise is the source of experts' autonomy and explain why this expertise is a valuable resource in two basic ways. First, asymmetries of knowledge and theoretical understanding make politicians highly dependent on experts' technical abilities. Second, higher levels of expertise also allow technocrats to act strategically in order to advance their policy interests and increase their autonomy.

Then, using four factors, I explain the variation in the strength of technical autonomy across policy sectors and within the same policy sector across time. Fundamental for technical autonomy and continuity, the first two relate to the politics of policy: the political cost to the incumbent of poor policy performance and a balanced, plural constellation of stakeholders with interests in a policy sector. The other two factors, related to the technical aspects of policy – the level of technical complexity and the degree of policy consensus among experts – when present, also contribute to technical autonomy. Finally, I explain how these factors account for technical entrenchment and technocratic demise.

The following chapters provide the empirical basis for these conclusions. Chapters 4 (Colombia) and 5 (Peru) show that economic policy is, in general, a more likely sector for the emergence and persistence of technocratic autonomy because of the presence of the aforementioned factors. In two of the episodes (Colombia 1958–2013) and Peru (1990–2013), economic experts enjoyed considerable autonomy and achieved long-lasting continuity. Nonetheless, these favorable conditions are not always present in economic policy. Politicians' lack of awareness of the cost of irresponsible economic policy, an unbalanced constellation of stakeholders, and a lack of expert consensus explain why Peruvian economic experts were dismissed from their posts during the early 1980s when they faced strong political and business opposition.

The three health policy episodes discussed in Chapters 6 (Colombia) and 7 (Peru) confirm that expertise is technocrats' most valuable resource for gaining and maintaining autonomy. The episodes also confirm that health experts' autonomy is lower and more vulnerable than is the autonomy achieved by economic experts. In two of the three episodes (Peru 1990–2002 and Colombia

1966–1970) politicians eventually dismissed experts, whereas in the Peruvian episode conditions were more favorable for the emergence and continuity of technocrats' autonomy. Variation in three of the aforementioned factors explains the more positive trajectory of health policy experts during the second Colombian episode (1993–2013). In this episode, by creating a single health care system, which included the formally employed and the poor, the reform enhanced the political cost of poor health policy performance. Also, the constellation of stakeholders in health policy became more plural and balanced, partly due to expert-driven policy reforms. Colombian experts were more successful at limiting the intrusion of other actors in health policy and have remained influential actors in Colombia, although their autonomy remains more vulnerable than in the economic cases. Finally, the health economics approach made health policy more complex, thus making it easier for experts to exclude other actors from policy making.

4

Economic Technocrats in Colombia (1958–2013)

> There currently exists a nucleus of young Colombians dedicated to their country's development above and beyond partisan considerations, often accepting positions with the government and sacrificing more lucrative opportunities in private employment. They are engineers, economists, agronomists, the new technocrats. (Robert Dix, 1967, 199)

Under the National Front pact (1958–1974), particularly during the presidency of Carlos Lleras Restrepo (1966–1970), economic technocrats in Colombia established themselves as the architects of their nation's economic policy. Since then, the Ministry of Finance (MF), the National Planning Department (NPD), and the Central Bank (*Banco de la República*) have become widely recognized for the technical excellence of their cadres. Colombian politicians also became informally constrained to appoint only highly qualified individuals selected from a small group of candidates to top positions in these institutions. Diverse scholars highlight the continuity and influence of this economic technocracy as a distinguishing feature of Colombian politics (Urrutia 1991; Hershberg 2006, 5; Thorp 1991; Thorp and Durand 1997; Edwards and Steiner 2008, 67; Mares 1993; Meisel Roca 1997; Juárez 1993a; 1993b; Leal Buitrago 1990, 35; Berry and Solaun 1980, 447–448).

In clear contrast to Colombia's Andean neighbors, technical cadres have consistently held economic positions for more than fifty years. The practice of appointing technocrats has persisted through varied political and economic conditions, including some hypothesized to be unfavorable to technocratic continuity. Experts were picked to head ministries under technically inclined presidents, such as Virgilio Barco (1986–1990), César Gaviria (1990–1994), and Juan Manuel Santos (2010–), but also under more "political" leaders such as César Turbay Ayala (1978–1982), Belisario Betancur (1982–1986), and Álvaro Uribe (2002–2010). Experts remained influential during periods of bipartisan rule from 1958 to 2002 but also under a hegemonic President, Álvaro Uribe in the 2000s. Furthermore, this autonomy remained high under good economic conditions in the early 1970s, the early 1990s, and late 2000s as well as during international economic crises in the 1960s, late 1970s, early 1980s, and late 1990s.

How did experts achieve autonomy and continuity in economic policy areas of the Colombian state? Why did economic technocrats in Colombia successfully entrench their expertise decades before those in other Latin American countries? What explains experts' domination of economic agencies while technocrats in other policy areas of the Colombian state remained weak and vulnerable to political interference? In what follows, I show how experts' deliberate actions to institutionalize technocratic control, and particularly, how the high scores in the four factors put forth in my theory explain economic experts' autonomy and entrenchment in the MF and NPD.

The chapter proceeds as follows. In the first section, I discuss the political and economic context in which autonomous experts rose to prominence in the Colombian state during the National Front (NF) and sketch the biographies of several top economic officials during the last decades in order to show technical continuity. In the following section I present evidence of experts' autonomy, highlighting how experts' preferences are frequently at odds with those of politicians and those of the private sector, the actors with the strongest interest in economic policy making. With this picture of continuity and autonomy clear, in the third section I show that economic experts deliberately pursued strategies that helped entrench their technical autonomy. Specifically, Colombian technocrats built their power through a gradual process of careful institution building. In this final section, I detail how the four factors highlighted by my theory explain this long-lasting autonomy of the Colombian technocracy.

TECHNOCRATIC EMERGENCE AND CONTINUITY

The emergence of an economic technocracy in Colombia can be traced to domestic and international conditions during the NF period when the liberal and conservative parties agreed to share power. The NF pact developed out of the two parties' mutual interest in ending the dictatorship of General Gustavo Rojas Pinilla (1953–1957), stopping the fierce partisan violence wracking the country, and fostering economic development. According to the agreement, the parties would alternate in power for a period of sixteen years (1958–1974), or four presidential terms, supporting the NF's presidential candidate in each election. To guarantee parity, all governmental positions (seats in Congress, governorships, ministries, and local authorities) were divided equally, and a constitutional provision stipulated a two-thirds majority in Congress for approving laws. Electoral competition against the Front's candidates was legal, but challengers had to run under liberal or conservative banners even if they lacked the official support of the parties. The pact, strongly supported by Colombian elites and approved in a referendum by 90 percent of the population, became part of the Colombian Constitution (Dix 1967; Payne 1968; Hartlyn 1988; Gutiérrez Sanín 2007).

To achieve its goals, the NF aimed to professionalize the politicized bureaucracy and rationalize budget allocations. Both parties recognized that patronage politics caused drastic changes in personnel every time power shifted between parties, leading to gross inefficiencies and partisan conflict (Payne 1968; Schmidt 1974; Gutiérrez Sanín 2007, ch. 2; Leal Buitrago and Dávila 1990; Dávila and Delgado 2002). The Pact of Sitges (July 1957), the "peace treaty" between the parties that served as the basis for the NF, proposed that administrative positions be filled by meritocratic appointment, with bureaucrats prohibited from intervening in political controversies, and sanctions for those who abused public office (Schmidt 1974, 431–436).

In an effort to fulfill these promises, liberal Alberto Lleras Camargo, the first NF president (1958–1962), undertook ambitious reforms. These changes included the creation of a newmerit-based civil service, institutions to enforce the neutrality of public employees, reforms in various state sectors, and a national school for high-level bureaucrats, the *Escuela Nacional de Administración*, modeled on the French *École Nationale Administratif* (interviews with Vidal Perdomo 2006 and J. Cárdenas 2008). Part of this reform effort focused on economic institutions. For the Front's development strategy to work, it was crucial to strengthen economic institutions, depoliticize budget expenditures, and professionalize policy making by appointing technocrats (Mares 1993; Juárez 1993a; 1993b).

The existing institutional configuration of economic policy making in Colombia limited the government's ability to manage the economy and respond to external crises. Among other problems, a strongly partisan Congress retained considerable control over the budget (Juárez 1993a, 76–78; Mares 1993, 462–463). Congressmen used these powers to restrain the executive's policies or provide their constituencies particularistic benefits like higher wages for workers or tax benefits for special interest groups (Mares 1993, 463–464; interview with Urrutia 2008). Additionally, the Central Bank board of directors, which set monetary policy, consisted of bankers and not state officials (Mares 1993, 460).

These economic reforms account for economic experts' entry into government. In an effort to control and rationalize public expenditures, President Lleras Camargo increased the power and technical capacity of the National Planning Office (later called National Planning Department, or NPD) (Caballero Argáez 2008; Urrutia 2008; Juárez 1995,103–113). Formally dependent on the presidency, the NPD, together with the MF, was given responsibility for the budgetary process. Two economists with postgraduate degrees from Harvard, Oxford, and the London School of Economics were appointed as NPD directors during Lleras's term. Another reform was the creation of a special council, the *Consejo Nacional de Política Económica y Planeación*, to coordinate development investment. Initially composed of two executive ministers and two members of Congress, the council later grew to include two technical advisers (Mares 1993; Juárez 1993a; 1993b).

In addition to the domestic reform push, international linkages and leverage also prompted politicians to appoint technocrats to state positions. Linkages with IFIs and foreign aid agencies led to efforts at technical rationalization of the budget allocation process.[1] Additionally, the Alliance for Progress, a U.S.-sponsored program to support democratic countries in Latin America, offered considerable funds to states joining the program if they presented a development plan. The NPD designed a ten-year plan to fulfill this requirement and obtain these funds (Botero 2005, 19).

More important than these linkages, external leverage increased the executives branch's need for economic expertise. Balance-of-payments crises in 1957, 1961, 1963, and 1965 forced the government to negotiate with the IMF in order to obtain new loans to balance the budget (Berry 1980, 289). Both Alberto Lleras and the NF's second president, Guillermo León Valencia (1962–1966), relied on economists with the necessary skills to negotiate with IFIs. The IMF demanded substantial devaluations to solve the crises, which in 1957 and 1962 threatened political stability. Pressure from IFIs caused the government to seek technical counter-proposals that only highly trained economic experts could prepare (Grindle and Thomas 1991, 5; Mares 1993; Botero 2005; interview with Botero 2009). Although these economic crises were by no means comparable to the hyperinflationary disasters of the eighties in Latin America, they did contribute to the recruitment of experts.

As a result, economic technocrats in the executive branch gained more control over economic policy during the NF. These changes happened incrementally, frequently by using, and abusing, the president's power to declare states of siege and emergency. Because of the 1962 crisis, for example, President Valencia reformed the planning council and changed its name to the *Consejo Nacional de Política Económica y Social*, or CONPES. The reform effectively excluded members of Congress from holding positions in this institution. Valencia also convinced Congress to create a Monetary Board composed only of members of the executive branch. In a pattern that became increasingly common, this new board also included two technical advisers (Mares 1993, 463).[2]

As discussed in detail later, these newly appointed experts not only helped the executive branch to achieve more control over the economy but also brought into the state a whole new set of ideas and an agenda that transcended immediate executive interests. Experts maneuvered to entrench their autonomy, orient

[1] The demand for a professional planning office was not new: as early as 1949, the World Bank's first international mission was sent to Colombia and recommended such a reform (Sandilands 1990).

[2] Law 21 of 1963. The Board consisted of the ministers of Finance, Development, Agriculture, the head of the Planning Department, the manager of the Bank of the Republic, and two permanent technical experts.

the country's development model outward, and diversify the economy through the promotion of nontraditional exports (Juárez 1993a, 467–468; 1993b, 25).[3] These reforms included the creation of an insulated technocratic institution, ProExpo, charged with financing and promoting these exports (Díaz-Alejandro 1973). To prevent its politicization, the institution's governing board was composed of business representatives and technical experts.

These technocratic trends became even more pronounced under President Carlos Lleras Restrepo (1966–1970), who further strengthened the NPD and other economic institutions. The NPD was tasked with authorizing public investment, administering external loans, and designing a national development plan at the outset of each administration. Additionally, Lleras reinforced CONPES to centralize development and economic policy decision making (Thorp 1991, 150). The NPD's director became the general secretary of CONPES, a move that considerably increased the NPD's control over public investment (Grindle and Thomas 1991, 83–87). In 1969, Lleras and his team also achieved a constitutional reform that further reduced Congress's role in economic policy making.[4]

Lleras surrounded himself with what is now identified as the first generation of Colombian technocrats: Economic Secretary of the Presidency Rodrigo Botero; Director of the NPD Édgar Gutiérrez; and Monetary Board advisers Hernando Gómez (1964–1968), Jorge Ruíz Lara (1965–1969), and Miguel Urrutia (1968–1970). Also under this administration, the technocracy took its first steps toward self-perpetuation as other young officials, such as Roberto Junguito, Guillermo Perry, and Carlos Caballero Argáez,[5] who are all now senior members of the Colombian technocracy, came to occupy leading positions in the NPD.

A particular event during these years constitutes a defining moment for the consolidation of economic expertise. In 1967, low international coffee prices caused another balance-of-payment crisis that led to a breakdown in negotiations between the government and the IMF. The Colombian government rejected the Fund's demands for a substantial devaluation (Juárez 1993b, 24; Díaz-Alejandro 1977, 184–206). In a televised speech, Lleras explained to Colombians that the "unfortunate differences" with the IMF were based on the institution's "anti-technical" demands (Thorp 1991, 148). The alternative was a solution proposed by Lleras's own experts: a "crawling peg" exchange rate that involved gradual devaluations. This move further increased the political importance of technocrats. Experts often remember it as the decisive moment

[3] Decree Law 444, March 1967.

[4] However, the reform was costly for the president. In exchange, congressmen received discretionary expenditure funds ("auxilios parlamentarios").

[5] For a report on the different generations of Colombian technocrats see *Cambio*, "Los Nuevos Gurús de la Economía," N. 816, February 2009.

TABLE 4.1. *Colombian Economic Officials (1958–2013)*

Ministers of Finance	Individuals (1)	Economists (BA or MA)	Graduate Studies in the US or Europe	PhD	Previous State Position in MF, NPD, or Central Bank	Worked for Business Before or After Appointment	Member of Political Party (2)
1958–1974	12	4	5	1	2	3	7
1974–1990	10	7	5	1	4	5	2
1990–2013	11	11	11	6	9	7	4
Directors of NPD							
1958–1974	10	7	7	1	5	3	3
1974–1990	9	6	8	1	8	5	1
1990–2013	13	11	12	7	11	7	1

Sources: M. Palacios 2001 and Meisel Roca 1997 and archival research.

1. Four individuals, José Antonio Ocampo, Eduardo Wiesner, Juan Carlos Echeverry, and Mauricio Cárdenas both held the position of Minister of Finance and Director of NPD. Roberto Junguito was Minister of Finance on two occasions (1983–1984; 2002–2004).
2. Formally, from 1958 to 1998 almost all individuals were sympathizers of one of the two traditional political parties. I count as politicians only those individuals who were party leaders or held elected positions at any level (local, congressional, or executive).

in which domestic experts could negotiate on equal terms with IFIs (interview with Botero 2009).

After Lleras's administration, a conflict emerged between experts in the NPD and the last of the NF governments under Conservative President Misael Pastrana. This conflict, discussed in more detail later, led to the resignation of several of Lleras's remaining technical appointees in the NPD. However, Pastrana did not reduce the technical qualifications of appointees to the NPD; a change in this technical team did not produce politicization in the NPD. This incident exemplifies how, even at such an early stage, incumbents were already informally constrained to appoint experts within economic institutions. By the end of the NF, liberal Alfonso López Michelsen (1974–1978) reappointed several members of Lleras's economic team to positions within economic institutions.

Since then, president after president has followed the precedent of making technical appointments to Colombia's economic institutions. The academic and professional credentials of top economic officials in the last five decades attest to this technocratic continuity. As Table 4.1 shows, the level of technical expertise within the government has increased consistently through time.

Economic advisers to the Monetary Board also possess this high level of expertise. I focus on these particular officials only until 1991, when constitutional reforms granted the *Banco de la República* institutional autonomy through the creation of an independent board of directors. As shown in Table 4.2, the

TABLE 4.2. *Economic Advisers to the Monetary Board (1958–1991)*

Economic Advisers to the Monetary Board	Individuals	Economists (BA or MA)	Graduate Studies in the US or Europe	PhD	Previous State Position in MF, NPD, or Central Bank
1962–1991	17	12	10	7	15

Sources: M. Palacios 2001, Meisel Roca 1997, and archival research.

board appointments have also been overwhelmingly technocratic, including several past NPD directors and ministers of finance (M. Palacios 2001).

Colombian technocrats have created their own recruitment process and informal career track. Commonly, they begin as economics students at the *Universidad de Los Andes*. Since 1970, thirteen of twenty-four NPD directors and eleven of twenty finance ministers held BAs in economics from *Los Andes*. In the 2000s, six of seven NPD directors and four of six finance ministers were undergraduates at this university. These young economists then have internships with state economic institutions or economic research centers such as the think tank FEDESARROLLO. Next, they hold intermediate positions in state economic institutions. Scholarships from these institutions allow these experts to obtain further specialized training at first-class foreign universities and then return to Colombia to assume higher state positions. Current Ministers Juan Carlos Echeverry (Minister of Finance) and Mauricio Santa María (Minister of Health) followed this career path. The stability of Colombian experts in economic institutions over the course of time is not in question. But is this continuity and informal career path a sign of autonomy? I explore this question in the next section,

TECHNICAL AUTONOMY IN COLOMBIA

Evidence of technical autonomy is that technocrats' preferences are distinguishable from – and often at odds with –those of politicians and economic actors in Colombia. To demonstrate this autonomy, I examine the argument that experts' stability and power in Colombia result from either patronage-based politicians or business actors and show how it does not, however, reflect the empirical evidence. In the process, I show how experts' preferences are often not aligned with those of these powerful actors.

Miguel Urrutia proposes that technocratic influence over Colombian economic policy depends on politicians' need to maintain patronage-based resources (Urrutia 1991). A characteristic of the Colombian political system, before and after the NF, is political parties' extensive use of patronage-based (Gutiérrez Sanín 2007; Hartlyn 1988; Wilde 1978). Urrutia argues that the absence of economic populism in Colombia is due to politicians' awareness that macroeconomic stability is necessary in order to maintain their patronage-based machines. Thus, "rational" politicians during the National Front chose to leave economic policy to experts. Although Urrutia acknowledges that technocrats are important actors in Colombia, his main factor explaining independent economic policy is the parties' decision not to interfere in the economy. The rejection of populist policies is paradoxically based in politicians' desire to protect a higher political interest: patronage-based resources.

Urrutia is correct to highlight that powerful, patronage-based political parties have restrained their intervention into economic policy and lived side by side with experts. However, in contrast to Urrutia, I argue that technocratic policy making is more a cause than a consequence of politicians' restrained intervention into economic policy. Two reasons make me reject the view that technocrats are ultimately dependent on politicians.

First, the argument that political moderation protects clientelistic resources is exaggerated. On the contrary, experts limit political control over clientelistic resources. Budgets and national development plans are first designed in the MF and NPD and then debated in Congress, which negotiates earmarks without making broad changes that affect fundamental macroeconomic lines. This planning process restrains politicians' access to patronage spending. As a result, the budget deficit remains under control despite frequent demands for additional spending. Even without formal protections to guarantee their independence, the NPD and the MF have counterbalanced strong clientelistic forces in Colombia that, if allowed to, would increase government expenditures.

My interviews with experts and politicians clearly document that these actors are often at odds. Economic experts are perceived as constantly limiting political expenditures, as confirmed by ex-finance ministers like Hommes and Junguito, and ex-director of the NPD and now Minister of Finance Echeverry. All these individuals consider negotiating skills to be crucial assets for Colombian finance ministers, whose jobs involve spending considerable time meeting with politicians to deal with their particularistic demands (interviews with Echeverry 2008, Hommes 2008, Junguito 2008, and Botero 2009). Experts keep Congress in check, opposing policies with potentially negative economic effects, and advancing technical policies through careful negotiation.

Political actors also share this view of opposing and counterbalancing forces. According to Senator Victor Renán Barco, one of the most powerful electoral barons in Colombian history and president of the Senate's budget commission for two decades, the relation between traditional politicians and technocrats is one of "tense cordiality." Experts and politicians see one another as limiting their preferences: experts are frustrated at not being able to drastically reduce clientelism, while politicians feel limited by experts' control over resources (interview with Barco 2006).

Additionally, Urrutia's theory assigns too much rationality to politicians. With few exceptions, politicians lack the technical skills that experts have. Politicians understand the political costs of economic crises but are less aware of the long-term consequences of particular policies or aggregate costs of budget expenditures, especially if the decisions are highly technical. Politicians constantly negotiate their demands with the finance minister and the *Dirección de Crédito Público* (Directorate of Public Credit), the MF office responsible for executing the budget. According to its current director, these negotiations are never ending, with politicians seeking to direct more funds toward their constituents and the ministry frequently blocking their demands (interview with Janna 2013). Crediting politicians with sufficient knowledge to make a tacit pact to not interfere in macroeconomic policy and restrain their budget appetite for clientelistic resources overstates politicians' foresight and understates experts' leverage.

The image of competing forces, with experts checking politicians who are demanding more resources, is closer to reality than Urrutia's rational theory. It is true that empowered technocrats and strong clientelistic political parties have coexisted for more than fifty years. Even though traditional parties have recently lost their dominance over politics without new national parties emerging, clientelistic regional and local politicians still remain highly influential. So this balance of power between politicians and economic experts persists.

Furthermore, an alternative view that presents experts as the agents of a president interested in restraining the clientelistic appetite of Congress would also not be accurate. As discussed in the following section, presidents certainly find experts to be useful officials that help to both keep Congress in check and advance reform goals. But technocratic economic policies also restrain incumbents; there are tangible limits constraining presidents' management of the economy. Experts are often the ones drawing the lines, sometimes even clashing with politicians' preferences. Turbay, Betancourt, and Uribe all disagreed with expert policies but nonetheless kept the economy under tight technical control. More broadly, some of the experts' development plans, such as moving toward a more open economic model with fewer industrial subsidies, were not part of the agenda of politicians such as Turbay or Betancourt. Reforms and changes adopted to achieve this goal were designed and gradually implemented through technical reforms. At times, incumbents' and technocrats' preferences coincided, as in the cases of Presidents Barco and Gaviria during the market reforms of the late eighties and early nineties. In other cases, however, experts pushed for reforms that transcended politicians' goals.

A second set of arguments to explain technical continuity in Colombia sees business actors as the real power behind experts. These arguments focus on structural forces and the influence of the private sector as crucial to experts' status. Estrada, for example, argues that the Colombian technocracy is closely related to dominant economic interests. During the aftermath of the NF, experts achieved a certain level of autonomy from local interests, but mostly for structural reasons (Estrada Álvarez 2005). This "relative" autonomy was based on the balanced conflict between industrial and export interests, which provided experts with enough room to gain some independence. This period was short-lived, however, as transnational interests and their domestic allies became dominant in the eighties. Since then, experts have become neoliberal in their approach. Similarly, Ahumada (1996) attributes market reforms conducted in the nineties in Colombia to the dominant position of transnational economic elites and their local allies.

Business interests are certainly important in Colombia and, as I argue later, have contributed to technical autonomy. Colombia boasts several business groups and well-organized private sector associations that constantly provide opinions about economic policy and thereby limit how much politicians can intervene in this policy arena. Economic policy consultation between the state and business associations is common. These associations enjoy substantial access to state officials, partly because of the revolving door between economic agencies and the business community (interviews with Echeverry 2008, Junguito 2008 and Montenegro 2008).

Nonetheless, these theories greatly exaggerate the power of economic forces over technical cadres. Technocrats are sufficiently distant from these actors to be considered as autonomous from them. To begin with, experts often hold different policy preferences than those of economic forces, which occasionally resulted in conflicts with certain powerful interests. In Colombia, experts faced coffee and industrial interests who sought to influence macroeconomic policy and block experts' attempts to tax their businesses (interview with Perry 2008). Before experts entered government during the NF, Colombia had pursued an import substitution industrialization (ISI) model that conceded extensive protective tariffs to industrialists. Additionally, coffee exporters received special benefits, such as a separate interest rate from that adopted by the *Banco de la República* (Juárez 1993a; 1993b; Mares 1993). Yet, as technocrats gained control over economic policy making, the development model became considerably more outward-oriented (Berry 1977, Juárez 1993b, 26; Mares 1993).

Additional evidence of technical autonomy is that even if there are close ties with business sectors, experts manage information selectively, choosing to not share some of it with business actors. This strategic behavior was evident in the response to the economic crisis in the early eighties (Thorp and Durand 1997, 223). The economic team led by Minister Roberto Junguito (1983–1984) was forced to significantly devalue the currency and then try to prevent inflation. The team achieved these goals through a reform plan that was kept secret from businesses and politicians outside the executive branch (Thorp 1991, 185–190; interview with Junguito 2008). Secrecy was necessary because these actions had negative short-term consequences for businesses. Had they known the plan, the private sector's reaction would have rendered the reforms ineffective. That the plans remained secret demonstrates that state experts were not subordinate to private sector interests.

More generally, a crucial problem with these theories is that the supposedly inevitable causal relationship of economic forces determining public policy does not hold in Colombia. The opposite is sometimes true: state policies, designed and implemented by experts, can modify the country's economic structure. As mentioned, since entering government, Colombian experts have usually supported export-oriented policies that slowly pushed the economy toward greater openness and less dependence on coffee exports. During the NF, these preferences led *técnicos* to clash with entrenched industrial and coffee interests. The incentives that experts introduced for nontraditional exports helped these new sectors gain considerably greater weight in the economy. After 1967, nontraditional exports grew impressively from 15 percent of exports in 1962 to 46 percent in 1972 (Botero 2005; Juárez 1993b, 27–28). Coffee gradually lost its predominance, falling from 67.8 percent of exports in 1963 to 52.2 percent in 1967 and a mere 20.9 percent in 1990. Thus, despite the lack of support from coffee and industrial interests for these policies (Juárez 1993a; 1993b; Mares 1993; Urrutia 1983, 42–43), experts induced significant structural changes in the Colombian economy.

Similarly, market reforms in the late eighties and especially in the early nineties were driven by experts who faced ambivalence and sometimes outright opposition from business associations and politicians. Reforms started at the end of President Virgilio Barco's administration (1986–1990) (Juárez 1993b; Juárez 1995, 156). Barco's economic team designed a market reform plan largely in secret. The government unveiled this plan by suddenly reducing 861 protective tariffs, a measure criticized by business leaders for lack of prior consultation.[6] Only afterward did economic officials meet with business associations to explain why the reforms were a necessity.

Market reform greatly accelerated and deepened during the government of César Gaviria (1990–1994) who, having served as Minister of Finance himself, shared policy preferences with economic experts. Under Minister Rudolf Hommes, public enterprises were privatized and the country opened to international trade (Edwards and Steiner 2008; Juárez 1995, 156–170). The government eliminated import licenses for all goods except some agricultural products and reduced tariffs from an average of 44.6 percent in 1989 to 9 percent in 1994 (Reina and Zuluaga 2006, 138; interview with Reina 2008). While some economic associations favored the reforms or accepted them as a necessity, others opposed these changes. However in general, business associations adopted a wait-and-see attitude (Cepeda Ulloa 1994; Juárez 1995; interview with Rettberg 2009).[7] These expert-designed reforms further altered the structure of

[6] *El Tiempo*, "La apertura económica: a mitad de camino," December 30th, 1990.

[7] *El Tiempo*, "Apoyo del nuevo gobierno reclaman gremios nariñenses," August 2nd, 1990; *El Tiempo:* "Gremios del agro arremeten contra la apertura," September 6th, 1990; *El Tiempo*, "Agro: apertura con protección," September 20th, 1990; "Creada comisión rural para defender intereses del campo," September 21st, 1990; "Contraescape," October 4th, 1990 (about criticisms of the reforms from small industrialists and commerce); "Pese a las dificultades, para los empresarios es irreversible la apertura: ya no se puede devolver," October 14th, 1990; "Agricultores y apertura," November 6th, 1990; "Apertura: los pies sobre la tierra," December 15th, 1990.

business interests (interviews with Junguito 2008 and Urrutia 2008). Theories about external pressures from IFIs can also be discarded because IFIs lacked the leverage to press for market reforms in a country with a stable economy.

As pointed out by Rudolf Hommes, the minister of finance who pushed these reforms forward during the Gaviria administration, Colombian experts were able to implement market reforms before economic conditions worsened, in clear contrast to other countries in the region (Cepeda Ulloa 1994; Edwards and Steiner 2008; interview with Hommes 2008). Experts shared the opinion that an early reform was preferable to waiting for economic conditions to deteriorate. That is, market reforms were part of a preemptive movement to make the economy more competitive and in line with experts' prescriptions; they were not the result of business pressures, political urgency, or external demands. Thus, contrary to economic-structuralist arguments, the direction of causality runs from experts to business rather than the other way around.

Moreover, in contrast to the structural and political interpretations of technocratic power, experts in Colombia have sought to insulate economic policy making from business and political interests. As mentioned, in the sixties technocrats successfully created a monetary board composed only of members of the executive branch, reducing private and congressional influence (Juárez 1993b, 23). Later, in the eighties, technocrats sought to further insulate the monetary board as well as reduce its power to provide special benefits to business interests and political elites (Botero 2005, 23). The 1991 Constitution gave this responsibility to an independent board in the Central Bank, as numerous economic experts had proposed (De Brigard 1995; Hommes 1988). The reform was designed by technical officials from the Central Bank and supported by Finance Minister Rudolf Hommes (Hommes 1995).

In sum, economic experts in Colombia cannot be considered either the product of enlightened politicians' choices or mere agents of the private sector. On the contrary, the evidence presented shows how economic experts have acted with significant autonomy in Colombian politics during the past fifty years. Experts have preferences and agendas that are not subordinate to those of other actors. Some academics and politicians suggest that bipartisan rule and the reformist intention of presidents during the NF is fundamental to explaining this technical independence in economic policy (Kline 1983, 105–107; interviews with Morales Benites 2008 and Llorente 2006). Although some of these actors were important reformists, these arguments forget that several of the NF's bipartisan reforms were never implemented or fell short of their goals. Large parts of the public bureaucracy remained unprofessional, and clientelism increased during these years (Leal Buitrago and Dávila 1990; Gutiérrez Sanín and Dávila 2000; Gutiérrez Sanín 2007, 146–147; Wilde 1978). Social reforms did not advance significantly (Hartlyn 1988, 112–118) nor did the National School of Management gain a reputation for providing high quality education (interview with J. Cárdenas 2008). Agronomists and health experts failed to achieve a similar entrenchment after the NF (interviews with Junguito

2008 and Ordóñez 2006). In contrast to their initial promise to modernize the Colombian state, the traditional parties learned how to share the spoils in most policy areas. What explains the exception of the economic technocracy?

EXPLAINING TECHNOCRATIC AUTONOMY IN COLOMBIA

My explanation of this exception is based on both technocrats' actions and, more fundamentally, the conditions surrounding economic policy in Colombia. On the one hand, experts purposefully maneuvered to enhance and entrench expertise in economic policy areas. At the same time, as discussed next, they benefited from high scores on the four factors – both pairs of politics of policy and technical aspects of policy factors – identified by my theory, which made economic policy a more likely area for technical autonomy to emerge and entrench.

Crafting Technocratic Autonomy in Colombia

As discussed, domestic and international factors contributed to the appointment of experts in Colombia. Once experts were in their posts, politicians gradually grew more dependent on technocrats in the Ministry of Finance, the NPD, and the *Banco de la República*. Technocrats boosted their relevance to politicians by providing contacts from IFIs, taking control of international negotiations, overseeing ongoing reforms, and designing development plans for new administrations (interview with Caballero Argáez 2007). Simply by occupying their positions and fulfilling their responsibilities, experts made politicians more reliant on technical advice.

However, experts also understood that profound changes were necessary in order to keep economic policy under their prudent technical management. The crisis conditions that led to experts' appointment were not dramatic enough to induce politicians to give them full control over policy. As a result, experts' efforts to craft lasting autonomy and guarantee technical policy in the long run became significant causes behind this outcome. In Colombia, experts deliberately used expertise as a political tool to achieve continuity. Rodrigo Botero, one of the first economic technocrats, calls the gradual entrenchment of technical expertise in the Colombian state a process of "intelligent design" (Botero 2005; 1995).

I identify two strategies through which Colombian experts established and maintained their autonomy in economic policy. First, *técnicos* made technical knowledge a requirement to participate in economic policy making in order to exclude less qualified individuals. Second, regarding institution building, experts adopted formal and informal rules to guarantee meritocratic technical recruitment in economic policy areas. Furthermore, they took steps to increase the size and influence of the epistemic community of economists within the state and Colombian society.

First, expertise proved a highly valuable resource for marginalizing other actors who previously exercised considerable influence over economic policy. Technocrats changed the terms upon which economic policy was debated and adopted in Colombia (interviews with Botero 2009 and Urrutia 2008). These experts were aware that their superior knowledge gave them an advantage that they actively exploited. Experts demanded that anyone questioning their proposals had to do so with mathematical or statistical arguments. The low level of technical skill found among Colombian politicians in those years made this strategy effective; economic expertise was particularly powerful because experts' opponents lacked it almost completely. Previous economic policymakers, usually lawyers linked to political parties and business groups, did not have similar technical abilities, which allowed experts to easily dismiss their arguments. During discussion of technical complexity details, this asymmetry proved crucial for excluding these actors and entrenching technical autonomy.

A second strategy experts used to entrench their autonomy was adopting institutions to guarantee the continued appointment of *técnicos* in economic policy areas and prevent politicization. Members of the economic team sought to institutionalize the NPD by developing informal networks with other experts in the bureaucracy, establishing linkages with political and economic elites, and reinforcing an internal meritocratic culture (Juárez 1995, 107; Grindle and Thomas 1991, 84–87). Experts in all other economic institutions followed similar strategies. To guarantee technical recruitment, the NPD, MF, and the *Banco de la República* adopted an informal rule: only officials with postgraduate studies in economics or related technical fields could be appointed to the higher ranks of these institutions. With very few exceptions, reaching the top position in the NPD unofficially requires, at a minimum, having engaged in doctoral-level studies (interview with Urrutia 2008; Urrutia 1991, 385).

Additionally, to ensure a continuous supply of candidates with adequate technical credentials and reinforce a meritocratic culture, Colombian economic institutions offered scholarships to their members to pursue graduate studies abroad. To receive these scholarships, candidates must have been accepted at a prestigious foreign university, which reduces the possibility of party-sponsored candidates. The *Banco de la República* started these programs in the seventies. The scholarships are actually loans, but they can be "paid" by working in the Colombian state for a number of years or contributing to academic research (e.g., publication in a refereed journal eliminates part of the loan). In 2005, the bank had twenty-one employees with PhDs in economics while another sixteen were abroad completing their doctoral studies (Botero 2005, 23). Similarly, since the sixties, support from foreign agencies has allowed the NPD to provide scholarships to its members. In 1986, this institution adopted a similar program to the one at the *Banco de la República,* although it focused on MA degrees (Juárez 1995, 108). Between 1986 and 2008, 215 NPD officials received scholarships for postgraduate studies abroad (Rentería 2008, 94).

Although many individuals who work for economic agencies could receive higher salaries in the private sector, these academic incentives and the prestige of working in highly reputable institutions has made these agencies attractive for qualified candidates (interviews with M. Cárdenas 2008, Echeverry 2008 and Santa María 2008).[8]

The first economic experts not only raised the technical sophistication of economic policy, but also sought to increase the broader legitimacy of their technical approach. They did so by supporting civil society institutions and promoting the study of economics. Colombia now has a strong "epistemic community" in economics with several journals and newspapers devoted to economic analysis. Technocrats write op-eds and actively participate in public debates. This community constitutes an important support group for Colombian experts and acts as a watchdog that limits political interference in economic policy. However, this community has not always possessed the strength that it does today. In earlier decades, not only did politicians and business people lack economic training, but national newspapers did not have economic analysts who could explain the costs and benefits of economic decisions to the general public (Botero 1995). The obvious paucity of economic expertise in the sixties motivated technocrats to work toward improving it.

Since the fifties, the Department of Economics at the *Universidad de los Andes*, especially its *Centro de Estudios sobre Desarrollo Económico* (CEDE) has become a first-class center of economic research and teaching. Thanks to linkages with international academic institutions that included special summer courses with world class academics in Bogotá, Los Andes has helped to provide a constant flow of Colombian students to PhD programs in the United States. Additionally, since the late sixties, state institutions such as the NPD and the *Banco de la República* have become centers of research. They publish economic journals and books focused on development, economics, and public policy, including articles in English to facilitate access to the most current research.[9]

More crucially, in 1970, Rodrigo Botero, with the support of business leaders and foreign agencies, created what became a highly regarded independent think tank and training center for the Colombian technocracy: *Fundación para la Educación Superior y el Desarrollo* (Fedesarrollo) (Botero 1995). Its creators hoped the center would produce reliable economic information that had been unavailable in Colombia at the time, conduct nonpartisan public policy analysis, train new generations of Colombian experts, and alert the public to

[8] According to Echeverry, these informal incentives attract very competitive candidates to the NPD. During his term in the NPD, he had eleven PhDs among his staff, even though their salaries were under $2,000 (U.S.) per month.

[9] Banco de la República has published *Revista del Banco de la República* since 1930 and, more recently, *Ensayos Sobre Política Económica*; *Planeación y Desarrollo* by the NPD has been published since 1969; *Desarrollo y Sociedad* from CEDE started in 1979. Eight universities in Colombia publish peer reviewed economic journals.

challenges facing the economy. Fedesarrollo's academic output is praised for its high technical standards and is seen as a model for technical institutions around the world (Gutiérrez Sanín 1997; Botero 2005; Braun et al. 2007). The same experts who initially develop proposals within the institution frequently go on to implement them as state officials.[10]

There are many examples of policy proposals that originated in Fedesarrollo. In the eighties, Fedesarrollo investigated the effect of market liberalization on Colombia and examined the difficulties of implementing it (Ocampo and Sarmiento 1987; Lora 1991; Amadeo 1993). In the nineties, the focus turned to the new institutions adopted in the 1991 Constitution and their problematic effects on long-term economic stability (Lora and Lanzetta 1995; interview with Santa María 2008 and Steiner 2009). This research raised concerns in the following years that led to reforms to rationalize public expenditures (Lora and Scartascini 2009). Fedesarrollo's influence on employment policies was also extensive (Braun et al. 2007, 14–20). Since the late nineties and especially during the last few years, Fedesarrollo has warned the government about the increasing costs of the health system for the public budget (Fedesarrollo 2009; 2010; interview with Santa María 2008).

In conclusion, Colombian economic technocrats acted strategically to achieve continuity in economic institutions. They have established procedures that guarantee their reproduction and created institutions outside the state to support technical economic management. These external institutions provide more career opportunities for experts, reducing their linkages to and dependence on the business sector which, as shown in the next chapter, is a weakness of Peruvian economic experts. Nonetheless, although important, experts' strategies are only part of the story. The conditions under which economic technocrats operated were more favorable than those for experts elsewhere in

[10] The following excerpt from a report about think tanks in Latin America shows this influence: "Many [top researchers and directors] have participated in the preparation and execution of economic and social policies. For example, many of Fedesarrollo's former Executive Directors have held offices of enormous influence not only in Colombia, but also internationally: Rodrigo Botero has been Minister of Finance of Colombia; Roberto Junguito rose on several occasions to the State, becoming Minister of Finance and Agriculture and board member of Colombia's Central Bank (*Banco de la República*); Miguel Urrutia became a Director of National Planning, Minister of Energy, and is still involved in policy issues as governor of *Banco de la República;* José Antonio Ocampo has been a former Minister of Finance, Minister of Agriculture, Director of National Planning and Executive Director of ECLAC (currently Undersecretary for economic Affairs of the United Nations); Guillermo Perry has been a former Minister of Energy, and Finance (currently he is the Chief Economist for Latin America and the Caribbean at the World Bank); Eduardo Lora became Senior Adviser of the Research Department at the Inter-American Development Bank and Juan José Echavarría gained entry in the *Banco de la República* where he became a board member. The actual executive director of Fedesarrollo, Mauricio Cárdenas, also held important positions in the Colombian administration where he served in several occasions as Minister of Economic Development (1994), Minister of Transportation (1998–1999) and Director of National Planning (1999–2000)" (Santiso and Whitehead 2006, 22–23).

the Colombian state. High scores in the four factors identified by my theory explain the success of Colombian technocrats in entrenching their autonomy.

Determinants of Technocratic Autonomy in Colombia

All four factors identified in my theory of technocratic autonomy are present in the case of Colombian economic experts. The first two politics of policy factors, politicians' awareness of the cost of bad economic policy performance and a plural and balanced constellation of stakeholders, are central to explaining technical entrenchment. These two factors, usually high in economic policy areas, limited political intervention and opened more space for experts to act autonomously. The latter two factors related to the technical aspects of policy, technical complexity and expert consensus, facilitated this process and are especially important in accounting for Colombia's early technical entrenchment compared to other Latin American countries. Over time, technocrats entrenched their power, and influential political and economic actors came to perceive them as responsible guarantors of stable economic policy. Incumbents are nowadays required to signal that they are responsible economic managers by making technical appointments.

As discussed in the first section, previous balance-of-payments crises meant Colombian politicians clearly understood the political cost of poor economic management when experts reached the state. After poor policy outcomes and violence in the previous decade eroded their legitimacy, party leaders in the sixties saw proper economic management as necessary to regain some prestige. Incumbent awareness of the political costs of an economic crisis insulated macroeconomic policy to a considerable degree. Turbulent international economic conditions further increased the political danger of irresponsible economic policy.

Certain episodes in particular illustrate how politicians are aware of the potential costs of economic mismanagement and seek to provide a range of guarantees that they will not politicize economic decision making. Evidence of such awareness comes from political campaigns in Colombia where candidates attempt to signal their commitment to technical continuity. For example, politicians may announce during campaigns who their economic advisers would be. Furthermore, incumbents quickly highlight the technical qualifications of their economic appointees and take great pains to show that they are choosing officials on technical grounds. Lastly, the recent first presidency of Álvaro Uribe exemplifies how even popular presidents attempt to gain technical credibility by respecting experts' decisions and not attempting to replace them with political appointees.

The election of President César Turbay Ayala in 1978 is one example that demonstrates politicians' awareness of the costs of irresponsible economic policy. Turbay, a long-time liberal regional leader, drew his power within the party from his regional support and patronage-based machine (Gutiérrez Sanín

2007, 133). Within the party, he opposed former President Carlos Lleras, who had strongly supported economic technocrats during his first administration, for the candidacy of the liberal party. As a result, Turbay's candidacy was perceived as a threat to technocratic continuity, a view that was furthered by supporters of Turbay who portrayed technocrats as *anti-políticos* that diminished politicians' control over important decisions (Gutiérrez Sanín 2007, ch. 3).

However, once on the campaign trail, Turbay sought to signal his commitment to fiscal responsibility. In particular, he accused his conservative rival of promising more than he could deliver and lacking economic realism.[11] After winning the election, he further showed his dedication to prudent economic management by appointing Eduardo Wiesner, a widely recognized expert who had served on his predecessors' economic team, as director of NPD (1978–1980). Wiesner later became Turbay's second minister of finance. Turbay's first finance minister also had a technocratic background.[12]

This case is not exceptional. All presidential candidates in Colombia bow to the technocracy on the campaign trail as Turbay did. Candidates campaign with an economic team, thus signaling they are committed to technical continuity and do not constitute a threat to responsible management. The leader of this team is usually a member of the Colombian economic technocracy. Former President Ernesto Samper (1994–1998) stressed that an effective team is crucial for conducting a successful campaign and gaining electoral credibility in Colombia because these individuals must present the candidate's plan to the citizenry, businesses, journalists, and other economic experts (interview with Samper 2008). Furthermore, the leader of the technical team is often appointed minister of finance following the election. When finance ministers step down during a president's term, the process of choosing their replacements is handled with similar care.

The election of Álvaro Uribe (2002–2010) highlights how even popular incumbents treat economic policy very carefully because of its potential political costs. In 2002, Uribe won an impressive 52 percent of the vote and ended 150 years of two-party rule. A liberal leader, Uribe decided to run as an independent when the party denied him the candidacy. Like his rivals, Uribe also appointed a well-reputed economic adviser in order to show his commitment to technical expertise. Alberto Carrasquilla, a PhD in economics and widely recognized expert, was Uribe's main economic adviser during the campaign.

After his election, Uribe and Carrasquilla met with three prestigious Colombian technocrats, former Finance Ministers Roberto Junguito, Rudolf Hommes, and Guillermo Perry, to discuss how to address the country's growing economic problems (interview with Perry 2008). Although he considered picking Carrasquilla, Uribe opted to appoint Junguito, an even more prestigious technocrat, as minister of finance to send a clear signal that the first

[11] *El Tiempo* "Betancur no sabe cuánto vale lo que ofrece: Turbay," May 4th, 1978.
[12] *El Tiempo*, "Habrá política económica sin sobresaltos," August 10th, 1978.

independent president in 150 years would respect technocratic continuity and fiscal responsibility. Similarly, Uribe selected Santiago Montenegro, also a reputed member of the Colombian technocracy, to head the NPD. As discussed below, at the height of his popularity and under better economic conditions, President Uribe did take some actions that clashed with economic experts and even appointed a less technically qualified minister in his second term. Nonetheless, he ultimately maintained a considerable level of technical expertise in these institutions.

These examples show the extreme awareness Colombian politicians have regarding the political cost of mismanaging the economy. Experts constitute a guarantee, or at least a signal, of proper management. These actions show that politicians perceive breaking the informal rule of technical appointment as politically risky.

Regarding my second factor, since the sixties there has been a plural and balanced constellation of stakeholders in Colombia. Political parties, business associations, and IFIs all have an interest in economic policy making. The power of these actors has varied over time, with some gaining more power and others losing it, but most of these plural forces have coexisted for the last six decades. All these actors closely follow economic decisions and expect technical picks during changes of ministers. Incumbents are aware that not meeting these expectations will produce a backlash from these actors. Experts profit from these cross-pressures that limit the control of any one actor over economic policy. *Técnicos* represent a guarantee to all actors that there will be no abrupt changes to the status quo. As a result, experts have more leeway to advance their policies and remain autonomous.

For most of the period under study, two political parties dominated Colombian politics, making it more difficult for incumbents to take control over economic policy areas. As mentioned, the political opposition closely monitors economic agencies. Profligate ministers constitute a risk to the oppositions' electoral chances. In Colombia, the opposition party traditionally kept a close eye on the incumbent to prevent the use of economic policy in the incumbent's favor.

An incident shortly after the end of Carlos Lleras' government that threatened technical continuity exemplifies the check each party held over the other. This incident also demonstrates the importance of maintaining high levels of technical expertise in economic institutions for Colombian presidents as early as the 1970s and the mechanisms through which stakeholders protected technical continuity. Conservative President Misael Pastrana (1970–1974), the last NF president, confirmed the appointment of Jorge Ruíz Lara, a former Monetary Board adviser and member of Lleras's technical team, to head the NPD. Some months later, however, Pastrana clashed with Lara and NPD officials over the priorities of the development plan approved by Lleras. Pastrana's actions were perceived as an effort to politicize the NPD, which led Lara to resign (Gutiérrez Sanín 1997; Schmidt 1974, 442–443).

Liberal leaders criticized Pastrana's decision for breaking the commitment to nonpartisan economic policy (interview with Perry 2008). Experts also expressed their dissatisfaction with the decision and warned about the costs of politicizing the NPD (Schmidt 1974, 442–443). Ultimately, those criticisms led Pastrana to appoint an expert, Roberto Arenas, as head of NPD. Although a conservative loyalist, Arenas had a technical education that guaranteed responsible, nonpartisan management of the NPD. The prestigious economist Lauchlin Currie became Pastrana and Arenas's adviser, a clear signal from the president that partisan use of the NPD would not be tolerated (interview with Arenas 2006). Even though several members of the departing economic team, as well as former President Lleras Restrepo himself, criticized the content of Pastrana's new development plan, its technical quality was not disputed (Arenas 1972, 3; CORP 1972; interview with Botero 2009). Liberal Alfonso López Michelsen won the next election (1974–1978) and rehired the experts that had left during Pastrana's government. As this conflict shows, powerful political actors constrained the incumbent to respect the informal rule of appointing technocrats to stop politicians from abusing macroeconomic policy. Thus, criticisms from liberal party leaders helped maintain a technical NPD in 1970.

Similarly, in Colombia there is a long history of IFI involvement in economic policy. IFIs are especially influential in periods of economic hardship, but there is continuous interaction with national governments even in good economic conditions. These institutions demand incumbents maintain a sound economic policy, which puts pressures on politicians to keep experts in their positions. As a result, experts' have more space to exercise their autonomy. The original appointment of experts is partially explained by pressure from IFIs. Experts were appointed to even out the technical disparity between IFI negotiations and the Colombian state (Botero 2005, 21). Negotiations with IFIs, and the need to respond technically to their demands, augment the space for technical autonomy.

Another factor facilitating the establishment of technocratic autonomy has been a fragmented private sector during those years that only became more atomized due to experts' promotion of nontraditional exports. Business associations are crucial stakeholders and exemplify how these cross-pressures restrain political intervention into economic policy and give experts more space. Colombia boasts several well-organized business associations, including powerful federations of industrial interests (*Asociación Nacional de Empresarios de Colombia*, ANDI), banks (*Asociación Bancaria y de Entidades Financieras de Colombia*, ASOBANCARIA), coffee growers (*Federación Nacional de Cafeteros*), financial institutions (*Asociación Nacional de Instituciones Financieras*, ANIF), and, more recently, an "association of associations" (*Consejo Gremial*) (Schneider 2004; Juárez 1993b, 17). These associations are often headed by former ministers of finance or NPD directors and produce highly technical reports on economic policy. In addition, these groups organize public forums where experts are invited to debate policy options (interviews with Echeverry 2008,

Junguito 2008 and Montenegro 2008). The opinions of these groups tend to be very technical and, even when coming from particular interests, are usually worded as promoting public welfare. By keeping a close eye on economic policy and making demands that frequently contrast with those of politicians and IFIs, business actors enhance experts' autonomy and make technical cadres in economic positions a necessity.

Uribe's government highlights the importance of stakeholders for restraining even popular and charismatic incumbents in the absence of strong political opposition. During Uribe's first administration, technical autonomy came under pressure. Minister Junguito lasted a year in his position.[13] According to the press, Uribe's style, as well as his tendency to constantly voice his opinion about economic policy, soon led to Junguito's resignation. The president, for example, publicly asked the Central Bank not to raise interest rates, which many saw as an attempt to influence an independent institution.[14] Junguito's resignation raised concerns among economic analysts who viewed him as a minister with enough clout to limit the president's power.[15]

Uribe appointed Alberto Carrasquilla, a minister with technical credentials, though less political aptitude. The press immediately voiced criticism from economic and political stakeholders who worried about this move's potential consequences for economic stability.[16] Nonetheless, Carrasquilla quickly showed that he was not Uribe's puppet. He opposed Uribe's intrusion into economic policy, as several clashes leaked to the public demonstrate.[17] In his second term (2006–2010), after winning reelection with an impressive 64 percent of the, thanks in part to significantly improved economic conditions, Uribe struck a stronger blow against economic expertise. For the first time in decades, the president chose a political appointee, Oscar Zuluaga, to head the MF. Although Zuluaga had an MA in economics and experience in business, he was a politician close to Uribe. Similarly, while the technical abilities of Uribe's second director of NPD were not in dispute, her credentials were less impressive than her predecessor's, and she also was considered close to the president.

Again, economic analysts in the press questioned the decision to name a more political minister of finance, calling it a risky break with tradition. The press also speculated about how this decision would be received by external stakeholders such as the WB, the IMF, and business investors.[18] Further, several

[13] *El Tiempo*, "Renunció Junguito," June 7th, 2003.

[14] The bank raised the interest rate anyway, disregarding the president's advice.

[15] *El Tiempo*, "Reacciones," June 7th, 2003.

[16] *El Tiempo*, "Editorial," June 8th and June 15th, 2003.

[17] *El Tiempo*, "Carrasquilla, en el ojo del huracán," November 27th, 2003 and "¿Ahora sí se va Carrasquilla?" January 31st, 2007.

[18] *El Tiempo*, "Uribe, a punto de optar por Minhacienda político," February 14th, 2007; "Oscar Iván Zuluaga, nuevo Minhacienda," February 17th, 2007; "Con Oscar Iván Zuluaga llega un político al Minhacienda y un escudero del gobierno al Congreso," February 18th, 2007.

senior technocrats publicly criticized Uribe during those years for bending the informal rules of economic policy making in Colombia (Botero 2010, 8; interviews with A. Gaviria 2008 and Kalmanovitz 2008).

Partly due to this scrutiny, the technical level of economic personnel remained fairly high under Uribe. Zuluaga kept technical vice ministers and did not profoundly change the MF's policy style (interview with Botero 2009). According to then Director of Fedesarrollo Roberto Steiner, Zuluaga in the end was "one of the boys"; he had economic training and was aware of the costs of politicizing economic policy (interview with Steiner 2009). Stakeholders, then, constituted an important limit to the politicization of the economy.

The technical aspects of policy factors also scored high for this economic episode, contributing to the high technical level of economic policy making. Regarding the complexity of technocratic knowledge, economic experts had a significant advantage over other actors. As discussed previously, technocrats used their expertise to increase the complexity of economic policy making. From then on, economic policy became a highly technical discipline, only accessible to those with expert mathematical, econometric, and statistical knowledge that vastly exceeded the capacity of others actors. This disparity made it easier for experts to take control of economic policy making.

As mentioned, the *Universidad de Los Andes* and CEDE, as well as foreign institutions, supported students undertaking postgraduate studies in economics at highly reputable universities. The idea was that these young PhDs would return to teach and conduct research in Colombian universities, but many ended up working in the new economic institutions (interviews with Junguito 2008; Perry 2008; Urrutia 2008). The new crop of *técnicos* brought with them econometric and mathematical skills that were rare in Colombia and distinguished them from the politicians and bureaucrats who had previously managed economic policy.

The new experts called their predecessors "*empíricos*," indicating that their expertise came from experience and not from real academic knowledge (Montenegro 2010, 16; Botero 2004, 17). Former Minister Rodrigo Botero recalls that it was common in cabinet meetings to hear nationalistic speeches justifying tariff protections or predicting dire consequences if experts' proposals were approved. The National Federation of Coffee Growers, for example, highlighted the importance of coffee to the Colombian economy in order to claim that the interests of coffee producers were the interests of the nation. Experts opposed these rhetorical claims with economic models and projections, calculated by Colombia's first computers (Botero 2004, 17). The projections showed that the alleged negative consequences of their proposals were much less severe than predicted, while showing that what benefitted coffee growers was not necessarily good for the nation.

Since then, economic policy debates in Colombia have only become more technically complex. According to Botero, this increase in the state's technical capabilities later prompted an increase in technical capabilities in the private sector as business associations' policy proposals now had to be defensible in technical terms (interview with Botero 2009; Botero 2004, 17).

This complexity also increases experts' leverage by making the consequences of policy decisions difficult for politicians to assess. Experts can thus dissuade politicians from pursuing certain policies by pointing out risks that they might not otherwise see. In the eighties, for example, the Ministry of Finance advised politicians not to borrow from international banks as many Latin American countries were doing (Thorp and Durand 1997, 223). Technocrats alerted the incumbent to the risks of taking out large loans and encouraged politicians to stand firm against international and domestic pressures (Botero 2005, 22; Botero 1976). The incumbent followed the experts' advice. Although criticized by politicians and business associations at the time, this decision helped spare Colombia from the effects of the debt crisis that produced economic disasters elsewhere in the region. In sum, the complexity of economic policy, especially given the scarcity of economic knowledge in Colombia, allowed experts to gain control over this policy area.

The second factor, consensus among experts, was also high. This factor is crucial in explaining the success of experts' strategies and the early emergence of a technocracy in Colombia. This fundamental agreement legitimates experts' favored policies among the epistemic community of Colombian economists, thus making it more difficult to justify any deviation from their policy prescriptions. Even if individual experts are fired, their replacements are likely to hold broadly similar views, thereby limiting the potential for major policy shifts.

The young Colombian experts who entered government in the late fifties and sixties shared a consensus about economic policy that derived from the core of economic ideas they had absorbed while studying abroad. This fundamental consensus advocated tight control of public debt, prudent monetary policy, market diversification through export-oriented development, and low tolerance for inflation (Meisel Roca 1997; Botero 2005, 21; Urrutia 2008, 72; interviews with Junguito 2008 and Perry 2008).

In line with these ideas, experts opposed what they considered negative and costly aspects of Colombia's ISI model as well as the economy's overdependence on coffee exports. Like their political superiors, experts disapproved of the institutional framework for policy making that limited executive power over economic policy but also saw many other reforms as necessary. Once they occupied influential positions, *técnicos* took gradual but firm steps toward reducing Colombia's dependence on coffee exports and strengthening public participation in areas of the economy where the state lacked a competitive advantage (e.g., in communications as well as in shipping ports and energy distribution) (Botero 2010, 7).

The level of consensus among experts during this period was remarkably high when compared to other Latin American countries where debates over economic policy were fierce. These debates, which the next chapter discusses in the context of Peru, were inspired by competing economic policy paradigms. On the contrary, Colombian experts debated economic management in technical terms while agreeing on some basic ideas of what constituted adequate policy.

Although during President Pastrana's administration (1970–1974) there were divergences among experts about the best policy options, these debates were based on some shared technical basis. The strong criticisms of Pastrana's development plan, for example, were made in highly technical terms (CORP 1972). Similarly, in the eighties, experts debated the "correct" degree of market liberalization but agreed on the overall thrust of policy reform (Ocampo and Sarmiento 1987; Lora 1991).

Some observers separate Colombian experts into various factions according to supposed different ideological positions. Experts displayed these particular differences in the debate during the eighties and nineties about the proper degree of liberalization of the economy (Estrada Álvarez 2005, 297–298). Also in 1990, during the process of enacting market reforms, Finance Minister Rudolf Hommes and his economic team clashed with Development Minister Ernesto Samper and his reputable technocratic advisers José Antonio Ocampo and Guillermo Perry about the adequate degree and timing of the trade opening (Flórez Enciso 2010).[19] However, according to two members of the Colombian technocracy who occupied opposite sides of this proposed ideological spectrum, these differences were rather moderate (interviews with Steiner 2009 and Perry 2008). Furthermore, when compared to the divisions between Peruvian economic experts in the early eighties, they appear even less significant.

As a result, Colombian politicians are limited to selecting their economic officials from a pool of experts that share a general policy consensus. Choosing officials from outside this group is easily criticized as irresponsible. As previously mentioned, when President Pastrana clashed with NPD officials, he appointed new experts that shared this broad consensus. Similarly, when President Alfonso López Michelsen (1974–1978) fired Minister Rodrigo Botero in 1976 after a series of conflicts with business interests, López appointed a previous minister of finance who guaranteed technical continuity.[20]

In sum, high scores in these four factors made economic policy agencies a more likely area for technical autonomy, leading to experts' entrenchment in the Colombian state. These factors explain the exception of the economic technocracy, where many other reform efforts to strengthen policy areas during the National Front failed.

CONCLUSION

This Colombian economic episode clearly shows the crafting of technical autonomy and its entrenchment over time. Experts' strategic actions, including

[19] *El Tiempo*, "Hommes vs. Samper. Primer Round," October 28th, 1990.

[20] *El Tiempo*, "Abdón Espinosa Valderrama, nuevo Ministro de Hacienda," December 18th, 1976; *El Espectador*, "No habrá timonazo económico: Min-Hacienda," January 17th, 1981.

the active exclusion of less qualified actors from policy making and adopting institutions to ensure their continuity, were important for achieving technical autonomy. Technocrats achieved this entrenchment even though Colombia did not suffer the severe economic crises that occurred elsewhere in Latin America and, as we will see, catapulted Peruvian experts to power. Colombian technocrats entrenched this technical autonomy step by step, by building an epistemic community within the state and civil society, which demonstrates that experts' strategic actions helped advance their autonomy.

The four factors my theory identifies as facilitating technocratic autonomy scored high for Colombia. My two politics of policy factors scored high, as is usually the case in economic policy. First, previous crises made incumbents aware of the costs of delivering poor economic policy. Additionally, a balanced constellation of powerful stakeholders increased the space experts had to adopt their own decisions while preventing backlashes against themselves.

Scores in the technical aspects of policy factors have also been high since the sixties in Colombia. The complexity of economic policy made it necessary for politicians to delegate economic management to experts and facilitated experts' exclusion of other actors from policy making. Experts' own work raised the level of technical knowledge required to participate in economic debates, which in turn contributed to greater technical sophistication among economic actors in Colombia. Lastly, the first group of experts shared a basic consensus of what constituted proper economic policy and successfully reproduced this consensus in subsequent generations of Colombian economists. This early consensus among experts helps explain the rise of *técnicos* in this country long before they arose in other Latin American countries. This consensus also limited the available options in economic policy. The contrast with early Peruvian technocrats, who disagreed strongly on policy, highlights the relevance of this factor for technocratic autonomy. A well-entrenched technocracy exists today in Colombia.

After Uribe left the presidency in 2010, the technocracy regained its supremacy over economic policy and other policy areas in the state. Juan Manuel Santos, Colombia's new president and former minister of finance, appointed reputed Colombian technocrats to head the NPD and MF as well as other sectors like health and agriculture. His first cabinet was considered as technocratic as those of Lleras Restrepo, López, or Gaviria. Technocratic power remains alive and well in Colombia.[21] As mentioned playfully by Rodrigo Botero, "In Colombia it is nowadays easier to predict who will be the next minister of finance than to guess who will be the next president" (interview with Botero 2009).

[21] *Semana* "Dream Team," July 7th, 2010.

5

Economic Technocrats in Peru (1980–1983; 1990–2013)

Economic technocrats in Peru entrenched their autonomy in the Ministry of Economics and Finance (MEF) during Alberto Fujimori's first administration (1990–1995). Appointed to fight hyperinflation and implement drastic market reforms, technocrats were successful in transforming the Peruvian economy according to their preferences and, in the process, gained considerable autonomy (Abusada et al. 2000; Cortázar 2006; Dargent 2008; Wise 2003). This technocratic autonomy has remained quite high under the more competitive democratic regime that was instituted after Fujimori's downfall in 2000 due to a corruption scandal. Even President Ollanta Humala (2011–), a strong critic of the "neoliberal" economic model associated with MEF technocrats, maintained this technocratic continuity in the ministry after his election.

Peru, however, also experienced an earlier episode in which technocrats failed to entrench themselves in the MEF. During Fernando Belaúnde's second administration (1980–1985), orthodox economic experts appointed to the MEF initially had significant control over economic policy. These experts, which the press dubbed "Team Dynamo," were insulated from political and business pressures in an effort to implement orthodox economic reforms designed to dismantle Peru's inefficient Import Substitution Industrialization Model (ISI) development model (Conaghan and Malloy 1994; Wise 1986; Webb 1994). Yet eventually, the economic team failed to counter criticisms from politicians and business leaders that questioned both the reforms and the imperious style of decision making. In 1983, Belaúnde fired these experts and a similarly strong technical team did not return to the MEF until 1990. I argue that lower scores in both of the politics of policy factors and in one of the technical aspects of policy factors from my theory explain this lack of entrenchment. The incumbent, and politicians in general, did not fully comprehend the political costs of poor policy performance, and the constellation of stakeholders in economic policy was unbalanced. Additionally, during the eighties in Peru, there was no technical consensus about what constituted adequate economic policy, which made unseating orthodox experts easier. The contrast between these two economic episodes supports my theory and highlights how the factors I propose contribute to technical autonomy and entrenchment.

The chapter proceeds as follows. The first section examines the first technocratic experience, highlighting how "Team Dynamo" faced unfavorable conditions for technical autonomy that eventually led to its dismissal. The second section discusses the rise of *técnicos* under President Fujimori and their continuity even after he left power. In this section, I demonstrate how technocrats were autonomous from politicians, by highlighting the restraints experts imposed on political actors and businesses, and by discussing the limitations of perspectives that see the private sector as the source of current technocratic power in the MEF. Although less concerned with institution building than are Colombian experts, I show in the third section that Peruvian technocrats have also acted strategically to insulate the MEF's autonomy. In the fourth section, I explain how the higher scores in the four factors of my theory explain entrenchment in this second economic episode.

ECONOMIC TECHNOCRATS IN THE EIGHTIES

After twelve years of military rule, Fernando Belaúnde (1980–1985), the president the armed forces had ousted in 1968, won the 1980 presidential election. Belaúnde appointed Manuel Ulloa, a lawyer and businessman, as his cabinet chief and minister of economy. Ulloa recruited a group of economic experts charged with reforming Peru's struggling ISI model (Webb 1994, 357). Adopted during Juan Velasco Alvarado's (1968–1975) revolutionary military government (Trimberger 1978; Stepan 1978: Lowenthal 1975), economic policy under ISI included increasing protective tariffs for local industry, nationalizing numerous foreign-owned firms especially in mining and oil, and creating multiple state corporations. The second phase of the military government, under the Presidency of Francisco Morales Bermúdez (1975–1980), had to deal with the high costs of this hastily imposed development model (Conaghan and Malloy 1994). Following successful negotiations with IFIs to obtain more international loans, the military government began to dismantle the ISI model (Mauceri 1995, 11).

These measures, however, faced stiff resistance from unions and political parties that demanded deepening, rather than dismantling, ISI. These groups were well represented in the 1978 Constituent Assembly and successfully pushed for the inclusion of substantial social and economic rights in the 1979 Constitution in an attempt to institutionalize ISI policies. When they took office after the 1980 election, Belaúnde and his *Acción Popular* (AP) party were not committed neoliberal reformers. Bealúnde believed that massive infrastructure projects would foster medium-term development that justified a public deficit in the short term. However, the president also understood that the only way to finance these projects was with international loans that were conditional on "putting the house in order" through IFI-backed neoliberal reforms (interview with Abusada 2010). The only clear domestic political support for an orthodox reform plan came from the *Partido Popular Cristiano* (PPC), a small right-wing political party in the governing coalition.

As a result, the new AP-PPC government faced a dilemma: orthodox economic reform seemed necessary to stabilize the economy but precluded politicians from following through on their campaign promises or delivering material goods to their constituencies. Belaúnde opted for a mixed strategy: insulate "Team Dynamo" in the MEF to allow them to implement neoliberal economic reforms, while developing infrastructure projects in the country in response to social and political demands. Belaúnde's preferences were never clearly aligned with those of his technical team, and tension between the two emerged in the following months.

The leader of the technical team was Roberto Abusada, a young expert appointed as vice minister of commerce in the MEF. Given Ulloa's responsibilities as cabinet chief, Abusada gained considerable control over the ministry's day-to-day operations (interviews with Abusada 2010, Hume 2007 and Arrieta 2007). Other members of the team serving as vice ministers and advisers in MEF were U.S.-trained economists, such as Jorge González Izquierdo and Brian Jensen, and professors of economics or engineering at the *Pontificia Universidad Católica del Perú* (interviews with Arrieta 2007 and Hume 2007).

From the time it was appointed, this economic team was quite insulated. Politicians and businessmen criticized experts for being too "theoretical" and lacking "practical" knowledge (Conaghan and Malloy 1994, 152–155; Mauceri 1995). Ulloa was the only contact between the president, his party, and the MEF. Due to this insulation, business groups' demands were mostly channeled through other members of the government, such as the minister of labor.

Experts designed a series of tax, financial, and commercial reforms to implement immediately after Belaúnde's inauguration in July 1980 (Webb 1994, 359–366). Two very sensitive political issues were included in this plan: technical control over foreign credit to finance public works and the privatization of state enterprises that were a source of political patronage (Pastor and Wise 1992, 87). Also, experts proposed incentives to promote foreign investment in agriculture and mining (Pastor and Wise 1992; Webb 1994). Even though nominally neoliberal, these reforms included what most contemporary economists would consider commonsensical measures, such as reducing the mounting public deficit, halting inflation, lowering costly subsidies of little impact, and privatizing some noncompetitive state enterprises (interview with Abusada 2010).

From the beginning, experts had a limited capacity to carry out their policy preferences. A first obstacle arose soon after experts arrived in the MEF.

President Belaúnde and *Acción Popular* requested the ministry delay reforms until after the November municipal elections because they believed removing price subsidies would hurt their electoral interests (Conaghan and Malloy 1994, 152; Webb 1994; interviews with Abusada 2010 and Hume 2007). The MEF agreed, but the planned reforms did eventually proceed following the elections. Several of these reforms were instituted using the president's broad emergency powers conferred in the 1979 Constitution to avoid congressional negotiations.

The team undeniably implemented some of its proposals, especially in international trade, which shows meaningful technical autonomy. Tariffs were reduced from an average of 66 percent to 32 percent in 1981 (Wise 1986, 27). Sticking to an orthodox agenda, the team even cut subsidies for nontraditional exports, a sector that could have served as an ally (Webb 1994, 364). The state's credit for manufacturers was drastically cut from 20 percent to 2 percent in 1982, and 98 percent of previously restricted goods could now be imported freely (Wise 1986, 27).

However, experts failed to achieve their goals in other areas, demonstrating that technical autonomy was limited in the early eighties when compared to the Colombian and later Peruvian cases. Very few state enterprises were privatized, none of which were major firms. Foreign investment did not arrive as expected, partly due to investors' lack of confidence that Peru would not make another statist turn (Pastor and Wise 1992, 88). In addition, although the MEF adopted regulations to limit international borrowing, it lost the battle to control the foreign loans used to finance infrastructure projects (Webb 1994, 361; Mauceri 1995, 12). The fiscal deficit even grew from 3.9 percent in 1980 to 6.7 percent in 1981. The debt crisis hit Peru in 1983, which placed more pressure on experts. Floods in northern Peru and attacks from the Shining Path, a radical Maoist guerrilla group, put the government under even more stress.

Partly due to these problems, Ulloa left government in January 1983. Abusada had resigned several months before. Former central banker Carlos Rodríguez Pastor was appointed to head the MEF while several members of Ulloa's team remained in their positions. However, dealing with the debt crisis consumed all of the new minister's time and brought reform efforts to a halt (interviews with Hume 2007 and Webb 2007; Webb 1994, 366–367). Although more technically trained than Ulloa, Rodríguez Pastor lacked his political experience and, after facing similar pressure from politicians, was soon fired. A political appointment was made to appease opposition from politicians and business. The new minister, a member of *Acción Popular*, increased import tariffs, adopted new industrial subsidies, and restored some export subsidies (Webb 1994, 367). Furthermore, in an effort to take control over tariffs away from the MEF, the vice ministry of commerce was transferred to the Ministry of Industry. Belaúnde's final minister was more technically oriented and made some serious efforts to restore fiscal discipline but lacked the political power to do so.

Why did experts fail to entrench the significant autonomy they had during the early eighties? Evidence of this autonomy is their high insulation and influence over policy making. Although they did not achieve the level of comprehensive reform they desired, technocrats did implement noteworthy changes to economic policy. Nonetheless, experts possessed autonomy only in areas where they shared preferences with politicians and other actors, which was not enough to achieve continuity.

I argue that three of the factors in my technocratic autonomy theory explain *técnicos'* downfall. Even though economic policy is an area where technical autonomy is expected, and entrenchment more likely, the politics of policy factors were not as favorable as they would be in later years or as they were in Colombia during the sixties. First, politicians were inexperienced after twelve years of military dictatorship. They saw the state-centered economic model as a source of clientistic resources but did not fully appreciate the political costs of its inefficiencies. The hyperinflationary crisis later in the decade taught them that lesson. Second, the absence of a balanced constellation of stakeholders limited the possibility of maintaining technocratic autonomy. Several actors that otherwise might have counterbalanced politicians' interests aligned with political actors in attacking experts. In addition, the lack of technical consensus on economic policy facilitated politicians' and business' opposition to MEF experts by providing alternatives to expert-sponsored policies.

A crucial factor that limited experts' autonomy was the lack of urgency among politicians to address the deteriorating economic conditions (Conaghan and Malloy 1994, 152–155). Politicians, including members of the incumbent AP, understood the political costs of poor economic policy performance but not the serious repercussions of maintaining their patronage privileges or the important economic consequences of a high public deficit. Furthermore, politicians were more focused on building (or rebuilding) their patronage linkages. Therefore, congressmen and members of the executive branch from AP pushed for higher expenditures. The public deficit continued to grow and public enterprises became more inefficient, but *políticos* kept demanding patronage resources and opposed experts seeking to reign in debt spending. The best example is the AP's request that experts delay launching the emergency reforms for four months until after local elections.

The constellation of stakeholders, usually more balanced in economic policy areas where business and IFI interests are at play, was unbalanced in Peru partly due to experts' confrontational strategies. The immediate cause of technocrats' demise was political and business interests' opposition to their reforms. Predictably, orthodox experts encountered strong opposition from leftist parties and labor unions. However, they also faced coordinated pressure from the incumbent's party and business associations. Only IFIs supported the reforms; all other actors opposed experts in a clearly unbalanced constellation of stakeholders.

To begin with, *Acción Popular*, especially a faction in Congress known as "*provincianos*," questioned the direction of the reforms. Javier Alva Orlandini, vice president (1980–1985) and congressional leader of AP in the Senate, was the main opponent of Ulloa's reforms.[1] María Jesús Hume, a member of the team and later vice minister of commerce, attributes this resistance from AP politicians to the politicians' vested interests in public works and especially public enterprises (interview with Hume 2007). Because public enterprises offered well-paid patronage jobs, politicians of all partisan affiliations stood to lose from reforms. Facing opposition from within his own party, Belaúnde, as mentioned, did not actively support his economic team.

Business associations also mounted a strong campaign against the experts. As one might expect, the most forceful criticism of Team Dynamo's reforms came from the National Society of Industries, which claimed that the trade liberalization was destroying local manufacturing (Conaghan and Malloy 1994,153; interview with Webb 2007). Unexpectedly, similar attacks came from the association of exporters, who were supposed to benefit from the reforms in the medium term. However, various measures, such as a large reduction in export incentives, negatively affected exporters in the short run and led them to strongly criticize the reforms (Conaghan and Malloy 1994, 116). Thus, experts antagonized even those business sectors that they had hoped would support their goals. Under these conditions, experts' only allies were the IFIs. Business stakeholders, who just a decade later would value technocrats as guarantors of economic stability, were deeply critical of them.

Turning to the effects of technical knowledge, the lack of consensus among Peruvian economists made MEF experts more vulnerable. Criticisms from politicians and businesses were based on the assessments of other economic experts, who proposed alternative approaches to fixing the worsening economic conditions. The debates among economists mirrored the strong ideological polarization that prevailed in Peru during those years. Technocrats urging market reforms were counterbalanced by other experts who saw more, rather than less, state ownership and protectionism as the solution. This lack of consensus is aptly described by the president of the Central Bank (1980–1985), Richard Webb:

> From the late 1970s through 1985, with the exception of part of 1984, the Presidents in power and their economic teams were pushing an adjustment and reform agenda that was perceived as radical, exaggerated, insensitive, and technically flawed by a majority of the public. Most academic and professional "expert" opinion kept up a continuous critical barrage. Although the motivation was partly political, and partly linked to interests groups, much of the criticism was sincere. (Webb 1994, 372–373)

[1] In *La República*, January 7th, 1982, Alva publicly criticized that "technocrats do not identify themselves with the "populist" ideals of *Acción Popular*" and accused them of "lacking contact with the people."

Parties' campaign platforms in the 1980 election demonstrate this strong polarization (Universidad del Pacífico 1979). These platforms presented an impressive range of divergent economic proposals. Leftist parties promoted the nationalization of enterprises to achieve economic independence and reduce the power of the ruling classes. Even in the 1985 and 1990 elections, the United Left's (*Izquierda Unida*) government plans, drawn up by the party's economic advisers, continued to show support for a decidedly non-neoliberal program of state-led development (Izquierda Unida 1985; 1990; see also Malpica 1985).[2] Centrist parties were considerably less radical, but also defended ISI subsidies and public ownership of certain enterprises. On the other side of the spectrum, small right-wing parties such as the PPC proposed market reforms similar to those advanced by MEF technocrats.

This lack of agreement on economic principles facilitated resistance to experts from politicians and business interests. By using the technical assessments of economists who disagreed with orthodox reforms, these actors made their own opposition appear less particularistic and more publicly oriented, thus undermining experts' claim that their measures were absolutely necessary. Alva Orlandini, for example, asked a group of economists in 1981 to write a report evaluating MEF policies and proposing alternative measures (Conaghan and Malloy 1994, 154).[3] Similarly, business groups that opposed reforms used alternative economic assessments to justify their positions. For example, in 1981 and 1982, business associations organized "high profile public conferences" where outside economists criticized governmental policies (Conaghan and Malloy 1994, 154). This disagreement between experts, combined with an unfavorable political context, led to experts failing to maintain their autonomy.

For the remainder of the decade, the MEF was weak and politicized. Alan García's administration (1985–1990) introduced a series of heterodox economic policies that the headstrong young president promoted as a new development model. To justify his turn to heterodoxy, García blamed the crisis on his predecessor's liberalization efforts. The group of heterodox economists that the new president appointed as his advisers opted to deepen ISI in order to solve the country's economic problems (Thorp 1991, 121–125; Weyland 2002a, 81–84). The economic team worked outside the MEF, while members of APRA, known as "*ministros secretarios*" for their subordination to the

[2] These differences increased as the economic crisis mounted. Evidence of how different the evaluations of the alleged causes and solutions of the crisis were in those years is found in Dancourt et al. (1990) where five economists present alternative solutions to the orthodox remedies against hyperinflation.

[3] When interviewed for this book, Alva claimed that the press's version, in which Alva was painted as a fierce opponent of Ulloa, was wildly exaggerated. His intention when hiring these economists was only to show experts that there were alternatives to the measures they promoted (interview with Alva Orlandini 2007).

president and strikingly limited technical credentials, were appointed to the ministries. García enacted economic policies that IFIs and some local economists described as suicidal (interview with Abugattás 2007). The success of García's program was dependent on the progressive activation of the economy, which never occurred (Thorp and Durand 1997, 227–228).

An unprecedented economic crisis exploded in late 1987, but not even this emergency motivated García to change course and give the MEF more control over the economy (interview with Abugattás 2007). García preferred to finish out his term without suffering the political costs of drastic market reforms. The availability of economic options justified his decision to avoid painful adjustments (Weyland 2002a, 87–91).

The crisis caused severe political damage to the president and incumbent party. Polls (*Apoyo Opinión y Mercado*) from those years show how the economic crisis demolished Garcia's image. During his first two years in office, the president enjoyed overwhelming popularity, with approval ratings between 82 percent and 67 percent. When the economic crisis began in June 1987, García's popularity stood at 52 percent. These numbers plunged in the following months, reaching the low teens in 1989 before climbing slightly to the low twenties when García left office in July 1990. Inflation gradually became the most important problem in Peru, above terrorism and unemployment, and pessimism about the economy drastically mounted among the population. In July 1988, 73 percent of Peruvians disapproved of the government's economic policy and 62 percent believed that the economic situation would only worsen in the future. In April 1989, 89 percent of respondents considered the economic situation as bad. Certainly, the activity of the Shining Path also hurt the president's approval rating. However, because the level of political violence remained constant throughout García's administration, it does not explain the sharp drop in his popularity.

As all these numbers show, the economic crisis imposed immediate and debilitating political costs on the incumbent and severely jeopardized APRA's electoral interests. APRA came in third in the 1990 election with only 22 percent of the vote, a far cry from the 50 percent it had won in 1985. The crisis taught the lesson that prudent economic management was crucial to political success and that not doing so was costly.

THE RISE AND CONSOLIDATION OF NEOLIBERAL EXPERTS (1990–2013)

A decade later, neoliberal experts succeeded where Team Dynamo failed. Since 1990, a technocracy has firmly established itself in the MEF. Peruvians hold widely ranging views of MEF technocrats. For some, this technocracy is responsible for two decades of economic stability and impressive growth. For others, experts represent a detached-from-reality group that is more concerned with business interests than the "real" needs of the people. Furthermore, some question the undemocratic character of experts who seemingly force politicians into straitjackets that impede them from fulfilling their campaign promises. What neither supporters nor opponents deny is, in contrast with previous periods, the power of experts in Peru.

Economic technocrats were appointed when newly elected President Alberto Fujimori (1990–2000) realized that he needed qualified economists to combat the hyperinflationary crisis inherited from Alan García. A political outsider who rose at the last minute to defeat favored right-wing candidate Mario Vargas Llosa in the 1990 elections, Fujimori lacked a government program and a political party from which to recruit *técnicos* (Mauceri 1995, Cortázar 2006, 15–17; 18–19; interview with Du Bois 2007).

During the campaign, Fujimori had opposed Vargas Llosa's proposed neoliberal stabilization plan. However, facing strong demands from IFIs and foreign governments once in office, the new president opted for a drastic set of orthodox reforms that were enacted shortly after his inauguration (Stokes 2001, 47–53; Weyland 2002a, 116–118). Fujimori's first two ministers of economy, Juan Carlos Hurtado Miller (July 1990–February 1991) and especially Carlos Boloña (February 1991–January 1993), recruited the economic experts who designed and implemented these measures. Usually appointed as consultants or advisers to the minister's cabinet, technocrats initiated ambitious market reforms that transformed the Peruvian state and the structure of the economy (Abusada et al., 2000; Conaghan 1998; Dargent 2008; Ugarte 2000, 410; Wise 2003).

The crisis allowed for a neoliberal technocracy to emerge and take almost total control over economic policy, as well as negotiations with IFIs. Experts perceived these severe crisis conditions as a unique window of opportunity to achieve the changes they had not achieved in the previous decade (interview with Du Bois 2007). They reduced tariffs drastically and moved toward an export-oriented economy through sweeping market reforms. These measures were adopted quickly, sometimes obstructing and contradicting one another (Weyland 2002a, 124), with dramatic effects on Peruvian society. Industrialists lost their leverage while, in the medium term, other business sectors such as exporters and mining firms became more powerful (Durand 2006). Experts were also put in charge of broad institutional reforms. They launched an ambitious plan to privatize state-owned enterprises (including banks), shrink the bureaucracy, reform the pension system, and overhaul commercial and tax law (Abusada et al. 2000; Weyland 2002a, Wise 2003).

Peruvian experts were not just following orders from IFIs. Although many elements followed IFI guidelines, these reforms reflected Peruvian experts' deep prior commitment to market reforms and their own ambitious agenda. Experts drew many of their ideas for dismantling the ISI model not from IFIs but rather from Vargas Llosa's 1990 government plan (designed in part by Roberto Abusada). Even stronger evidence that experts acted independently of international influences is the fact that their reforms were more radical than those promoted by IFIs (interviews with Du Bois 2007, Hume 2007 and Abusada 2010).

The political insulation of Boloña's team only increased after the 1992 self-coup when Fujimori closed the Peruvian Congress. Without political opposition, the MEF pushed forward several additional reforms by presidential decree, including a radical plan to privatize public enterprises (Mauceri 1995, 26). In contrast to the early 1980s, most state enterprises were privatized. The subsequent dominance of Fujimori's supporters in the new Congress increased the economic team's capacity to implement their reforms. Since then, an annual poll has ranked the minister of economics among the ten most powerful individuals in the country, showing the importance these officials gained during this market reform period.[4]

Nonetheless, when the crisis began to subside in 1993, Fujimori dismissed Boloña, allegedly because of the business sectors' dislike of his confrontational style. As in the eighties, business groups complained to the president because they viewed the technocrats as too insulated from their interests. However, as the next section discusses, important stakeholders such as IFIs and business interests made it very clear that they did not want the president appointing a political lackey to head the MEF. While business actors disliked an insulated minister, at the same time these actors demanded clear signals that the economy would be kept insulated from political pressures. IFIs were concerned about the continuity of reforms. Fujimori chose Jorge Camet (January 1993–June 1998), an engineer who had been minister of industry and president of CONFIEP (the encompassing business association), as a guarantee that economic stability would not be jeopardized. To signal his commitment to a technical approach, Camet kept Boloña's vice ministers as well as part of his technical team with two important additions – Roberto Abusada and World Bank official Iván Rodríguez and former Minister of Industry (1983–1984)– who had been members of Belaúnde's first economic team and now served as Camet's advisers.

With the crisis receding, MEF experts realized that their political leverage was diminishing. Although privatization continued, the breakneck pace of reforms had slowed (Weyland 2002a, 162; interviews with Du Bois 2007 and Rivera 2011). The MEF focused on guaranteeing macroeconomic stability and controlling the public deficit, although they accepted that they could not fully reign in military spending or funding for social programs that maintained Fujimori's popularity (interviews with Du Bois 2007 and M. Ugarte 2006).

The line between technical autonomy and "division of labor" blurred half way through Fujimori's second term (1995–2000). Fujimori's commitment to technical autonomy weakened as he pursued reelection in 2000. Tensions emerged between *técnicos* on the one hand and the president and his military allies on the other. Starting in 1997, the military ministers began to voice their

[4] This ranking, previously prepared by Debate, is now prepared by *Semana Económica*, an influential economic magazine. http://elpoderenelperu.com/poderosos/1998

opinions during cabinet meetings on economic issues that were supposed to be under the purview of technocrats (interview with M. Ugarte 2006). These tensions led to the dismissal of Camet in June 1998. Although his successor, Jorge Baca Campodónico, had a more technocratic personal background, he lacked Camet's political weight in the cabinet and was dismissed in January 1999. Baca faced the beginning of an international crisis and was not able to build good relations with business interests, who strongly criticized him.

Appointing Víctor Joy Way, a legislator from Fujimori's party and former president of Congress, signaled that the president wanted more political control over economic policy to ensure his reelection to a third, although unconstitutional, term. As a result, the main group of technocrats that had been in the MEF since the early nineties left, and those who remained recognized that it had lost some of its political insulation (interviews with Du Bois 2007, M. Ugarte 2006 and Zamalloa 2007).

Nonetheless, the risk of politicization was short lived. The 1999 international economic crisis made Fujimori retreat and dismiss Joy Way because the potential cost of poor economic performance in an election year was too high. In October 1999, businessman Isaac Goldenberg replaced Joy Way. A year later, Fujimori went as far as to bring back Carlos Boloña in order to recover some economic credibility lost during the campaign. Boloña announced drastic spending reductions and more privatizations in an effort to put the economic program back on track.[5]

Shortly after winning reelection, Fujimori resigned from the presidency due to a corruption scandal. An opposition-led transitional government was installed to organize elections for early 2001. The caretaker government also began investigations into corruption within Fujimori's administration, including among some MEF experts. These investigations not only showed that experts had pragmatically coexisted with mounting corruption in the state but also led to the prosecution of some former MEF officials for allowing and allegedly benefitting from illicit activities during those years (Congreso de la República del Perú 2001).

Although these events harmed the MEF's prestige, the demand for technical policy making remained high. The transition government sent a clear signal of continuity by tapping Javier Silva Ruete, an economist who had been minister from 1978 to 1980, to head the MEF. Since then, neoliberal technocrats have remained in charge of economic policy in Peru, despite the transition to a more competitive democratic environment.

As Table 5.1 shows, the MEF has not only maintained its expertise but the level of technical expertise among economic ministers increased in the 2000s. During Alejandro Toledo's administration (2001–2006), all ministers were

[5] *El Comercio*, "Presidente Fujimori Anuncia Reducción del Gasto Público," August 15th, 2000. The expenditure exceeded by $436 million (U.S. million) the amount in the budget for the first six months of the year; *El Comercio*, "Cambios en Rumbo Económico," August 20th, 2000.

TABLE 5.1. *Economic Ministers in Peru (1980–2013)*

Ministers of Economy	Individuals(1)	Economists (BA or MA)	Graduate Studies in the US or Europe	PhD	Previous State Positions in MEF or Central Bank	Worked for Business Before or After Appointment	Member of Incumbent Political Party
July 1980–1990	10	5	2	1	3	4	9
July 1990–2000	6	2	3	2	1	5	1
July 2000–2013	10	9	10	6	9	9	0

1. Carlos Boloña (1991–1992; 2000), Javier Silva Ruete (2000–2001; 2002–2003), Pedro Pablo Kuczynski (2001–2002; 2004–2005), and Luis Carranza (2006–2008; 2009–2011) were ministers on two occasions.

highly technocratic. Similarly, all four of Alan García's (2006–2011) ministers of economy were independent technocrats – that is, the opposite of his politically subordinate ministers of the 1980s. García's first minister of economics, Luis Carranza, was a former vice finance minister in Toledo's administration. Even President Ollanta Humala (2011–), who made a failed presidential bid in 2006 as a strongly "anti-neoliberal" candidate and who still criticized economic management during the 2011 campaign, maintained technical continuity in the MEF. Humala appointed García's last vice minister of finance as his first minister of economy.

During these years, macroeconomic policy remained insulated from political intrusion, showing experts' clear autonomy from political actors. The end of hyperinflation in 1991–1993 did not bring the downfall of Peruvian technocracy that some commentators had expected (Conaghan 1998, 160).[6] On the contrary, experts have effectively continued to restrain political expenditures even during good economic times (2003–2011). According to political-agents theories, economic abundance should lead to political pressures for increased expenditure. Certainly, this pressure has existed. The impressive economic boom of recent years has led congressmen, regional presidents (elected regional authorities), local authorities, and even the president and ministers to publicly demand that the MEF loosen restraint on expenditures. Despite this pressure, the ministry and its experts have successfully held the line on public spending and effectively reduced politicians' discretionary expenditures.

Certain examples show this general pattern of tight economic control and autonomy from political interests. Technocrats in the MEF have implemented special programs to prevent patronage spending such as the *Sistema Nacional*

[6] In Conaghan's view, the Peruvian case and the events following Boloña's dismissal raise questions about "any overdrawn images about the power and autonomy of technocrats in Latin America" (Conaghan 1998, 160).

de Inversión Pública (National System of Public Investment: SNIP). Beginning in the late nineties, SNIP required publicly funded projects to gain prior approval from the MEF. This control produced strong criticism from regional and local authorities who alleged that the MEF was unreasonably stringent. More interestingly, incumbents also complained about such control. Presidents Toledo and García frequently criticized SNIP for not allowing them to fulfill their electoral promises, but nonetheless kept the system. Experts working in SNIP mentioned that the support of the minister of the economy and the vice ministers has been crucial for protecting their work from these external pressures (interviews with Von Hesse 2007 and Prialé 2007).[7]

In addition to SNIP, the MEF also developed a national financial system called the *Sistema Integrado de Administración Financiera* (SIAF) to control budget expenditures. Both systems were originally intended to apply only to executive ministries but have progressively been adopted in other state agencies as well as regional and local governments. The hallmark of both systems is that they reduce politicians' latitude to use resources (interviews with Barletti 2007, Mori 2007, and Guerrero 2007). These reforms have strengthened the technical management of the budget and centralized control over state expenditures, thereby clearly affecting political interests.

Similarly, MEF experts opposed Toledo's desire to create a cash transfer program in Peru similar to the ones already existing in Brazil or Mexico. MEF experts had divergent opinions about the need and benefits of a cash transfer program, but they agreed on an important point: thorough reforms to existing social programs were necessary before launching such an ambitious program (interview with Arrunátegui 2013). Eventually, the program was launched (*Programa Juntos*) but limited to some rural districts. The program has gradually grown during García's term, and especially under Humala's administration, but under close technical supervision to prevent its politicization.

Another example of experts' control over political spending is the limitation the MEF imposed on public expenditures before the 2006 election. Late in 2005, President Toledo publicly promised to not use state resources to influence the election in the upcoming year, without mentioning any concrete measures. Following the president's statement, his cabinet chief Pedro Pablo Kuczynski and the Minister of Economy Fernando Zavala announced that no more than 60 percent of the executive budget would be spent in the first seven months of the year. However, this decision was not thoroughly discussed with Toledo; these ministers simply seized the opportunity and headed off possible patronage demands from the incumbent party (interview with Kuczynski 2007).

As all these examples show, experts have effectively restrained political interests. This technocratic control did not come easily and involved frequent

[7] This account is confirmed by Toledo's former head of cabinet, Carlos Ferrero (2003–2005). Ferrero witnessed how Toledo accepted the advice of his ministers and backed down from his original requests (interview with Ferrero 2007).

clashes with the incumbent political parties, *Perú Posible* and APRA. Conflict during Toledo's term, for example, led to open debates in the media between MEF experts and members *of Perú Posible*, including congressmen and an aggressive vice president (Tanaka 2005, 30–31).

This technical independence does not mean that policy decisions have always conformed to the preferences of *técnicos*. As in Colombia, experts have had to negotiate and sometimes accept limitations to their influence. For example, several public sector salary increases Toledo promised during the 2001 campaign were approved against the experts' advice. Alan García also achieved some exemptions from SNIP procedures for public works. Yet overall, MEF experts have considerable control over macroeconomic policy and state expenditures, which is a sharp contrast from the spiraling deficits and irresponsible spending policies that prevailed during the 1980s. Thus, it is safe to say that, in general, MEF technocrats have achieved high levels of technocratic autonomy from politicians.

A more disputed issue is whether the MEF is autonomous from business interests. Some authors attribute experts' considerable leverage in Peruvian policy making to the power of business actors. For these authors, there is no doubt that Peruvian reformers reconfigured business power in Peru in the early nineties. Experts, aided by the crisis, dismantled the old economic model and reduced the influence of the then-powerful business sectors that benefited from it (Arce 2006; Durand 2006). However, there has since been less conflict between the MEF and the "winners" of market reform, especially financial, mining, and some export interests. This congruence has led political commentators and some scholars to argue that these new private interests have captured state agencies and forced politicians to appoint experts who support their interests. In this view, experts design policy and exercise their veto power for the benefit of these business groups (Durand 2006, 203–204).

For these authors, evidence of this capture is that experts are frequently hired from the business sector and then return to it after their appointments (Arce 2005; Durand 2006, 197–198). My research confirms that Peruvian experts' career paths are closely linked to private business, even more so than in Colombia where more technical career alternatives exist in universities and think tanks. With the exception of the Central Bank, and some think tanks (such as *Centro de Investigación de la Universidad del Pacífico* (CIUP) or the *Grupo de Analisis para el Desarrollo* (GRADE)), there are few job positions in Peru for technocrats outside of state economic ministries and the private sector.

Moreover, even experts recognize the capacity of these newly strengthened business groups to lobby for policy changes or oppose reforms they dislike (interviews with Boza 2007, Burneo 2007, and Waldo Mendoza 2010). Pedro Pablo Kuczynski, who has held positions in investment banks, enterprise boards, and mining companies, regarded claims that these interests had captured the state as exaggerated but agreed that the main weakness of economic policy making in Peru is the state's dependence on the private sector for qualified technical personnel. For example, when he tried to hire an expert in foreign debt to negotiate pre-payments, he had to look to investment banks for competent professionals (interview with Kuczynski 2007). The state did not have officials with these skills, even in its more technical areas.

Because market reforms were designed to benefit certain economic activities in accordance with *técnicos*' wishes, certainly, overlap between the preferences of technocrats and these new business actors is expected. In a way, the emergence of influential private sector actors demonstrates Peruvian experts' triumph over the business interests that defeated them in the eighties: experts changed the composition of businesses to fit their preferences. Yet, it is not inconceivable that these new interests could have captured economy agencies. However, if this were the case, we would not expect technocrats to contradict the particular interests of these businesses. In fact, my research shows that Peruvian economic experts do pursue policies against the wishes of these groups, thus demonstrating their independence.

First, macroeconomic policy making remains quite insulated and oriented toward a free-market approach despite the opposition of some business sectors. The continuation of market reforms is a result of technocratic, rather than private sector, preferences. Evidence of this is that MEF decisions still take business unawares, such as in 2007 and 2010 when the MEF reduced protective tariffs to the surprise of businesses (interview with Carranza 2010).

Second, there are recent cases in which Peruvian technocrats have clashed with business interests, showing that these predominantly harmonious relationships are not exempt from conflict. For example, experts have rejected demands from powerful business actors, such as textile producers, to adopt protective tariffs for their products.[8] Additionally, Peruvian experts during the last decade sought to extract revenues from economic activities that benefited from market reforms. In particular, the tax authority, SUNAT, and MEF adopted a financial tax in 2003 despite stiff resistance from banks. More recently, these agencies have imposed a tax on stock market revenues, which the financial sector strongly opposed.[9] MEF experts also participated in the negotiations between President Humala's government and mining enterprises to raise taxes on this sector. These negotiations produced a substantial increase in revenue from mining companies.[10] Another point of disagreement is infrastructure projects such as ports and highways, which the MEF largely refuses to fund even though these improvements are important to business interests and are demanded constantly.

Finally, Ollanta Humala, a candidate private sector elites strongly opposed during the 2006 and 2011 campaigns, also appointed so called "pro-business" technocrats to key economic positions. These included Humala's first Minister of Economics Luis Miguel Castilla, whose background is very similar to his

[8] *Gestión*, "No Habrá Salvaguardias a Hilados de la India," August 17th, 2009.

[9] *Gestión*, "Ganacias en Bolsa Esta Vez Sí Pagarán Impuesto a la Renta,"August 21st, 2009.

[10] *La República*, "Así Se Negoció con Las Mineras," August 28th, 2011.

predecessors. Humala's choices demonstrate that technical appointments do not occur simply because businesses have captured the state. As discussed later, caution and pressure from a variety of powerful actors, including businesses, politicians, and IFIs, informally constrain presidents to appoint experts. As discussed in my theory, the private sector is part of the explanation behind technical power and continuity, particularly in Peru, but it is only part of the story.

Thus, since 1990 economic technocrats have remained in charge of economic policy in Peru. They were powerful not only under a non-democratic government that insulated them from political pressure, but also during more competitive democratic times, when their political superiors are weak, do not possess congressional majorities, and do not necessarily have a deep commitment to orthodox economic policy. Experts in the MEF enjoy significant autonomy from politicians and business actors. In contrast to the situation of the 1980s, when Team Dynamo failed to achieve its policy goals, technocrats have managed to implement their macroeconomic preferences. What explains this success?

EXPLAINING ECONOMIC TECHNOCRATIC AUTONOMY IN PERU

The severity of Peru's hyperinflationary crisis gave experts immense control over the economy, which allowed them to take considerable steps toward adopting their market reform preferences (Weyland 2002a, 198–199). For two years, Peruvian experts enjoyed almost total control over economic policy. This delegation of power allowed technocrats to increase expertise in MEF, transform the state, and break down traditional sources of opposition to market reforms, overriding any resistance from political or business interests. By doing so, experts dramatically raised the technical level of policy making in economic areas and created a more favorable environment for their own survival. Because of this initial period, Peruvian technocrats did not need to be as strategic or invest as much energy in institution building as did their Colombian counterparts in order to entrench their autonomy.

When the crisis abated in 1993, experts lost some of their leverage. Nonetheless, the four factors identified by my theory of technocratic autonomy favored technical continuity and entrenchment. From then until now, a strong technocracy has remained in control of economic policy. In this section, I first detail some of the experts' strategies to build their autonomy and adjust to the more competitive political environment following Fujimori's fall. I then explain how the four factors of my theory explain this outcome and highlight some contrasts with the 1980s.

MEF Experts' Strategic Actions

Throughout these two decades, experts strategically advanced their preferences and protected their autonomy. Experts were deliberate in systematically advancing their goals, and these actions contributed to their growing control over economic policy making. To begin with, knowing that they had a unique opportunity to pursue market reforms even though politicians were not committed to them, experts rushed to implement these new policies (interview with Du Bois 2007). MEF officials used international negotiations and the leverage of IFI officials to break down domestic political resistance to reforms. Aware that President Fujimori was not willing to reform the pension system, for example, MEF experts asked IFI officials to stipulate those changes as part of their loan conditionality (interviews with Du Bois 2007 and M. Ugarte 2006). To make these changes irreversible, experts aimed to insulate parts of the state from politicians by creating new agencies or restructuring old ones. Besides increasing their capacity and adapting them to the new market economy, experts provided these institutions with formal protections against political pressures (Abusada et al. 2000, 13–62; Wise 2003, ch. 6, Cortázar 2006, 16–50; interviews with Álvarez Rodrich 2007, Escaffi 2007, Salazar 2007, Távara 2007, and Boza 2007).

When the crisis eased in 1993, experts became more strategic. Since 1993, experts have worried about patronage demands from politicians. With a dominant party in power, experts received informal protection and political support for their proposals. At the same time, however, Fujimori and his political allies also threatened technical autonomy, since there was no opposition party strong enough to resist them. Experts in the MEF realized that without an ongoing crisis, they needed new strategies to prevent political interventions into policy making (interview with Du Bois 2007). One of these strategies included opening an office in Congress so that economic experts could quickly coordinate with legislators. The office also collected information about economic proposals in Congress so that experts could take preemptive actions against such bills if necessary.

These strategic actions became even more important after the fall of Fujimori's government in 2000. With several parties sharing power in Congress, the experience of the eighties raised significant concerns within the MEF about political intervention. Nonetheless, experts adjusted well to competitive democracy. Expertise became a key resource for diverting political and socioeconomic pressures and advancing technocrats' preferred economic policies in this new political environment.

The MEF took preemptive actions to limit political intrusion in economic policy. To deal with a fragmented legislature, transition minister Silva Ruete strengthened the MEF's office of parliamentary coordination to maintain close relations with political groups and gather information about legislation under discussion in more than twenty congressional committees (interviews with Cáceres 2007 and Nakasone 2007). If a bill that the MEF opposed came onto a committee's agenda, this office would provide information and technical analysis to prevent its approval. If Congress still approved the project, the MEF would try to convince the president to veto the law. Similarly, the MEF consistently opposed congressmen's efforts to include special budget items targeting funds at their constituencies.

A similar preemptive strategy was also used against other sectors of the executive branch, especially ministries that spent funds in social issues, infrastructure and so forth. All projects coming before the Council of Ministers first went through a council of vice ministers. The MEF frequently used this preliminary stage to learn what policies the other ministries sought to implement and, if necessary, to oppose them (interviews with Nakasone 2007 and Vásquez 2007). The MEF, then, has developed a strategy to block measures of which experts disapprove.

MEF officials also coordinate with other actors to advance their proposals, sometimes even against the interests of the incumbent. During Toledo's administration, experts frequently cooperated with members of APRA, the most vociferous opposition party, which was regarded within the MEF as a more structured ally that had longer time horizons than did the disorganized *Perú Posible* (interviews with Cáceres 2007, Zavala 2007 and Zamalloa 2007). Under the APRA government, the rightist party *Unidad Nacional* worked closely with the MEF on several issues (interview with Lombardi 2008 and Morón 2013).

Off the record, two former high-ranking MEF officials said that experts frequently went to the press as strategy to deflect political threats. Bills that would clearly increase the budget or benefit a special interest group were analyzed in the ministry and, when necessary, leaked to the press. The MEF used this strategy, for example, to push for the reform of a costly pension law (Law 20530) that provided expansive retirement benefits to high officials, including members of Congress from the incumbent party.

Nevertheless, as in the Colombian case, experts' strategies do not fully explain this outcome. Politicians restrained themselves from intruding in experts' work, and several powerful actors made it very clear that they demanded technical continuity in the MEF. Furthermore, even if debates about the economy remain acrimonious, few economists question the basic aspects of macroeconomic policy and the benefits of an outward oriented development model. There was less opportunity for abrupt expert replacement. In clear contrast to health policy or economic policy during the eighties, Peruvian experts since the nineties, like their Colombian counterparts, have enjoyed favorable conditions to obtain and entrench autonomy. What were these conditions?

Determinants of Technical Entrenchment in the MEF

Although, as mentioned, economic policy is already a favorable area for technical autonomy, the four factors my theory identifies as advantageous to technocratic autonomy explain why experts successfully established independence beginning in the nineties. Due to the economic crisis, three of those factors had higher scores than were present in the past, with the complexity of economic policy remaining fairly constant since the eighties.

The two politics of policy factors scored high. First, as mentioned, the economic crisis and its disastrous consequences for Alan García showed politicians, especially the incumbent, the negative political consequences of poor economic performance. Incumbents learned that fiscal equilibrium and good economic outcomes were important for their interests. Additionally, the population became sensitive to signs of economic irresponsibility, and a significant number of citizens started demanding economic moderation. The crisis came up repeatedly in my interviews with experts and politicians as something that "no quieres repetir" ("you do not want to do again"). Actually, two former presidents of the Council of Ministers used this same phrase to describe why appointing experts to signal responsible economic management is important (interviews with Ferrero 2007 and Del Castillo 2007). As such, politicians are well aware of the high risks of making mistakes in economic policy.

Certain examples throughout these years show politicians' worries about the consequences of politicizing economic policy. Despite Fujimori's populist turn during his later years in office as he sought his second reelection, he was aware that inflation and bad economic indicators would negatively affect his popularity, especially since part of his political capital came from having solved the economic crisis. As a result, once reelected in 2000, he brought back Economy Minister Boloña in an effort to show his commitment to responsible economic management through technical continuity.

Fujimori's successors have had even less room for irresponsible economic management, as they lacked similar economic credibility. Presidents Paniagua, Toledo, García, and Humala have all bowed to technocracy in order to subdue fears of irresponsible policy making. In the cases of García and Humala, this fear was especially high due to the uncertainties in the international economy following the 2008 financial crisis. All these presidents presented their ministerial picks as credible commitments of economic responsibility both in front of the population and stakeholders. As mentioned in the introduction, Toledo complained about technocrats frequently, but continued appointing them to the MEF. His commitment to technical continuity was already more firmly established, since he had campaigned in 2000 against Fujimori while promising to maintain the economic model. During his first term, he appointed Pedro Pablo Kuczynski to establish his commitment to technical and economic continuity. As discussed later, García and Humala made similar picks in order to leave no doubts about their commitment to technical continuity.

The potential consequences are visible when presidents do not clearly signal that they plan to continue with existing economic policy. An example that shows this fear of reprisal from citizens and organized stakeholders was Paniagua's appointment of Javier Silva Ruete, a reputed technocrat, as minister of economy. In his inauguration speech, Paniagua announced before Congress his commitment to economic stability. However, he included an ambiguous phrase that his government would "analyze the structure of our foreign debt" and "negotiate multilaterally to adjust these payments" accordingly.[11]

[11] Mensaje del Presidente Constitucional del Perú Valentín Paniagua Corazao November 22nd, 2000. http://www.congreso.gob.pe/museo/mensajes/Mensaje-2000-2.pdf

The declaration was perceived as a potential sign of instability. Stakeholders voiced their concerns immediately, demanding more precise information about this policy. Two days later, Silva Ruete appeared on a radio show to address these concerns, clarifying that the president's words by no means indicated a change in Peru's relationship with IFIs.[12] This evaluation of foreign debt never occurred.

The extent that higher awareness of the political costs of poor economic policy limits political intrusion should not be overstated. Individual politicians, especially in Congress, continue demanding patronage resources that experts still must keep in check, showing that "political learning" alone does not explain technical stability (interview with Zavala 2007 and Carranza 2010). Even so, greater awareness of the costs of poor economic outcomes certainly contributes to increased technical autonomy.

Second, a more balanced constellation of powerful stakeholders developed during the nineties. IFIs and business interests closely monitor economic policy and perceive experts as a safeguard against irresponsible policy making. These stakeholders follow economic appointments in the MEF. Business associations have become more sensitive to the costs of politicized economic policy (interview with Morales 2007), which limits the incumbent's control over economic policy making. Even industrialists, who are frequently strong critics of MEF technocrats ("*no se ensucian los zapatos*") and their neoliberal policies, recognize that technical management is preferable to political abuse of the economy (interview with Farah 2007).

This close vigilance can be seen in moments of ministerial replacement. These episodes highlight how once experts have gained prestige, stakeholders ranging from the political opposition, to businesses and IFIs, reinforce the demand for preserving technocracy and preventing politicization. In the Peruvian case, the fear of a return to the politicized economic policy of the eighties elevated these concerns.

The importance of stakeholders became clear during the days after Boloña was fired in 1993. Stakeholders made their opposition to political appointments in the MEF very clear. Press reports from those days show how the presence of experts had become a guarantee against this outcome. Business associations, IFIs, and even political leaders demanded that Fujimori clearly signal his commitment to economic continuity.[13] *El Comercio*, the most influential newspaper in the country, demanded that technocrats continue to manage economic policy through the MEF in order to guarantee stability.[14]

Similarly, Paniagua, Toledo, García, and especially Humala were also informally pressured by stakeholders to appoint technocrats in the MEF and

[12] *El Comercio*, "Se Reducirá el Presupuesto en Donde Sea Posible," November 25th, 2000.

[13] *El Comercio*, "Cotización de la Deuda Peruana Cayó con la Renuncia de Boloña," January 6th, 1993; "Se Afectó Limpieza de Moras con el BM y el FMI por Renuncia de Boloña," January 8th, 1993.

[14] *El Comercio*, "Editorial," January 6th, 1993.

maintain economic stability. These presidents lacked Fujimori's economic credibility among these stakeholders and thus had to earn their trust by maintaining prudent economic policy. Both García and Humala appointed former MEF vice ministers in an effort to signal credibility to stakeholders. Former populist Alan García responded to stakeholders' worries of a second economic disaster by appointing a highly reputed *técnico* – Luis Carranza – as economic minister.[15] A former vice finance minister and then investment banker, Carranza provided the signal García required to show he had broken with his first government's policies. The government presented the appointment as a commitment to economic stability (interview with Del Castillo 2007). Along with naming the minister of foreign affairs, Carranza's appointment was one of the first announcements García made upon taking office, and it was positively received by diverse stakeholders including IFIs, external investors, and business actors.[16]

For his part, Humala reappointed the central bank president to his post and promoted APRA's vice minister of finance, Luis Miguel Castilla, to head the MEF. These moves served to assure stakeholders of Humala's commitment to maintaining economic policy, given that Castilla was clearly aligned with the demands of business interests and IFIs. As in the case of García's first minister, stakeholders reacted positively to the decision, voicing their support in the media.[17]

Castilla has become a symbol of economic stability and a guarantee that Humala will not return to the left-leaning economic proposals he espoused during the 2006 and early 2011 campaigns. The director of one of Peru's most important business magazines described Castilla as the only effective "guarantee" against Humala's populist tendencies.[18] The minister has remained in his position for more than two years, which is the longest tenure for an economy minister since Camet in the nineties. Every time rumors surface that Castilla may resign because of a conflict within the executive branch, or simply because of exhaustion, the press voices the demands of powerful stakeholders that he, too, respond. Since taking office, Castilla has had to deny his resignation at least four times in order to squash rumors and calm stakeholders.

The short tenure of Víctor Joy Way in 1999 shows how stakeholders react to nontechnical appointments. Analysts, the political opposition, and some

[15] *Bloomberg Report*, "Peruvian Bonds Climb to Nine-Week High on Carranza Appointment," July 21st, 2006. Similarly, *The Financial Times*, "Garcia's Choice of Finance Minister Cherished," July 21st, 2006.

[16] *Gestión*, "Alan García Oficializó Nombramiento de Luis Carranza Como Ministro de Economía, July 21st, 2006; *Gestión*, "Banca de Inversión y Standard and Poor's Saludan Decisión," July 21st, 2006; *Gestión*, "Empresarios Confían en que el Nuevo Ministro Mantendrá en Orden Cuentas Fiscales," July 21st, 2006.

[17] *La República*, "CONFIEP y ADEX Confían en Crecer," July 23rd, 2011.

[18] *Semana Económica*, Editor's Opinion, "El Garante Castilla y la Precaria Confianza," June 24th, 2013.

sectors of the business community all criticized Joy Way's politically driven nomination.[19] To counteract these criticisms, Joy Way quickly tried to signal his commitment to technical policy making. He announced that he had hired international consultant Jeffrey Sachs as his adviser, although Sachs actually took part in only a few meetings, and presented a group of prestigious local economists as part of his new economic team.[20] However, these signs were insufficient in the context of an international crisis, and Fujimori soon dismissed Joy Way as conditions abroad worsened.

In addition, technical aspects of policy factors also ranked high in Peru, thereby contributing to technical continuity and entrenchment. The complexity of economic policy explains both why politicians delegated so much control over economic policy during the crisis and then had difficulty recovering it after the crisis receded. As discussed, the sophistication of economic policy in times of crisis explains politicians' willingness to surrender it entirely to experts. Lacking expertise in economics, most politicians have had little ability to design and implement reforms, which they instead leave to technocrats.

From then on, experts insisted that policy making comply with higher technical standards, thus tilting the playing field in their favor. The MEF's reports to Congress are written in a technical language, usually highlighting the costs of legislative bills. For example, supporters of reforming Law 20530 used highly technical evaluations from the MEF to show the allegedly high costs of the existing pension system. Furthermore, SNIP requires technical evaluations to assess the cost and sustainability of investment projects. Thus, technical complexity increases experts' leverage over political and socioeconomic actors in Peru.

Lastly, the technical consensus about what constitutes adequate economic policy was considerably higher in the nineties than it was in the eighties. The hyperinflationary crisis conclusively discredited "heterodoxy" and other populist alternatives to "prudent" economic policy. Economists now agreed that economic stability and low inflation must be cornerstones of economic policy (Leiteritz 2010, 137–138). This stronger consensus can be seen in the opposition to experts' policies. Although leftist economists question the prevailing market model, often severely, they now are significantly more moderate and have much more in common with their neoliberal counterparts than they did during the eighties (Leiteritz 2010, 133–134).

[19] *El Comercio*, "Ante los Cambios del MEF Empresarios y Analistas Coinciden en Exigir que no Se Opte por Medidad Populistas," 6th January 1999. *La República*, "Oposición: Será un Gabinete Electorero," January 4th, 1999.

[20] *La República*, "Joy Way Asegura que Política Económica No Será Modificada," January 7th, 1999, and "Joy Way Anuncia al Súper Equipo," January 7th, 1999; *Gestión* "Se Mantiene el Modelo," January 7th, 1999. *El Comercio*, "El Economista Jeffrey Sachs Asesorará al Ministro Joy Way," September 1st, 1999, and "Joy Way Presentó a Su Nuevo Equipo de Asesores Económicos," January 10th, 1999.

Government plans demonstrate such a growing consensus around macroeconomic policy. The economic proposals of the anti-market candidate in 2006 and 2011, Ollanta Humala, were more similar to the dominant paradigm than the heterodox proposals of the eighties (Partido Nacionalista 2006; Gana Perú 2011). The diversity of technical viewpoints, then, significantly shrank after the crisis and the consensus that emerged has remained strong (interviews with Burneo 2007 and Morón 2007).

In conclusion, the total delegation of power into the hands of experts in 1990, experts' careful maneuvering, and the suitable conditions in economic policy thereafter allowed experts to entrench their autonomy. For more than two decades, these conditions have remained favorable, thus raising the likelihood of technical continuity.

CONCLUSION

This chapter shows how Peruvian economic experts failed to achieve their goals in the eighties but subsequently rose and successfully entrenched after 1990. The change of conditions between decades explains this variation. The hyperinflationary collapse gave experts considerable power, enabling them to control economic policy, and silenced critics of their market-based approach, facilitating the emergence of expert consensus. Additionally, a more balanced constellation of stakeholders emerged in the country while politicians became more aware of the costs of delivering poor policy outcomes. Stakeholders such as business, IFIs, and the political opposition, came to see management by experts as preferable to leaving politicians in charge of economic policy. As a result, economic experts became autonomous actors in Peru.

These episodes reinforce the conclusions of the Colombian case. Expertise is the most important resource for economists in order to make politicians more dependent on them and to insulate their policies from political actors, business groups, and IFIs. Apart from the context factors, these cases confirm the relevance of the two technical aspects of policy factors. The technical complexity of economic policy increases the importance of expertise and limits external intrusion. The relevance of expert consensus for technical autonomy, present in Colombia since the sixties and in Peru since the nineties, is also confirmed. At the same time, the first Peruvian case shows how the lack of such a consensus can harm experts.

Some contrasts between the country cases also offer valuable information. Under non-crisis conditions in Colombia, experts made slower progress and faced greater challenges achieving autonomy. The dramatic crisis in Peru allowed neoliberal experts to quickly take control of economic policy. Colombian technocrats' strategies required more negotiation and institution building, possibly because the more deliberate pace of entrenching expertise made technocrats more concerned with institution building. With the exception of the Central Bank, economic experts in Peru have not adopted similar strategies to foster new generations of technocrats and thereby institutionalize technical control. Although there are some prestigious research institutes in civil society, such as the *Centro de Investigación de la Universidad del Pacífico* (CIUP) or the *Grupo de Analisis para el Desarrollo* (GRADE), such spaces are less prevalent than in Colombia.

The lack of this conscious institution building means the Peruvian technocracy lacks similar institutional support outside the state, which makes these experts more dependent on private sector future employment. Such proximity may be costly for technical expertise in the event of a crisis. Other actors can more easily question their legitimacy as technocrats, who in general would be associated with particular interests. In sum, Colombian experts seem better insulated to resist political and economic crises due to their conscious process of institution building that limits their proximity to business.

Additionally important, President Fujimori provided Peruvian experts significant insulation that increased their leverage. However, while *técnicos*' benefitted from the protection of a hegemonic leader, they also faced challenges arising from the president's reelection objectives. In more competitive democratic conditions, Peruvian experts have learned to act strategically and preemptively to maintain their autonomy, perhaps losing some of their leverage but gaining more independence from incumbents. As the Colombian case confirms, a strong political opposition constitutes one more relevant stakeholder that provides more space to economic experts and limits the incumbents' intrusion in economic policy making. In some policy areas, in which there are no strong stakeholders, the only guarantee of technical continuity may be political pluralism, which, as seen in the Colombian case, may not be enough.

The contrast between these economic cases and the following health policy cases allow me to further highlight the importance of the causal factors identified in my theory of technical autonomy. While health experts have also achieved significant autonomy from politicians, the less favorable conditions found in this policy area limit their degree of independence and make their successful entrenchment difficult. Technical aspects of policy factors, and more importantly, politics of policy factors, explain this strong divergence. As Chapters 6 and 7 show, health policy experts remain more vulnerable than economic *técnicos*.

6

Health Technocrats in Colombia (1966–1970; 1993–2013)

Health policy experts in Colombia achieved autonomy in two periods of the last sixty years. The first occurred under the National Front (NF) (1958–1974), when health policy experts conducted an ambitious modernization and expansion of the country's health care system. This reform effort reached its pinnacle under Health Minister Antonio Ordóñez, who served during the administration of President Carlos Lleras (1966–1970). However, conditions were not favorable for entrenching technical autonomy. Following the NF period, experts lost control of the health care system to politicians, who exploited it for patronage. Low scores in each of my theory's four factors show why health policy experts could not withstand this political intrusion.

The second period of autonomy started in 1993, when Minister of Health Juan Luis Londoño and his "Team Change" implemented a drastic overhaul of the health care system (González-Rossetti and Ramírez 2000; Ramírez 2004). Although experts lost some leverage after Londoño left the ministry, they remained influential either as ministerial advisers or in special offices within the ministry. More recently, during the government of Juan Manuel Santos (2010–2014), experts regained salience with three prestigious technocrats occupying ministerial positions. Even if not as stable as the economic technocracy, health policy experts have achieved continuity in Colombia and technical appointments are now expected in the Ministry of Health.

I explain the emergence of long-term technical autonomy in the less likely area of health policy with high scores in three of my factors. These scores rose in the decades between reform episodes, accounting for experts' entrenchment in the nineties but not in the seventies. Both "politics of policy" factors were clearly stronger in the second episode. First, enrolling most of the population in a single public system that has considerable financial problems made health policy a much more salient political issue, and the cost of poor policy performance rose. As a result, politicians needed experts to manage health policy in order to promote better performance. Second, a more balanced constellation of stakeholders emerged around health policy. On the international front, IFIs and other actors became increasingly interested in health policy reform during the nineties. Domestically, the reform itself created new stakeholders by fostering new groups of health care insurers and providers. The cross-pressures of these actors granted experts greater space in which to develop their autonomy. Together, greater political interest in effective health policy and a wider range of stakeholders explain a considerable degree of experts' increased influence in Colombian health policy.

Furthermore, higher policy complexity, one of the technical aspects of policy factors in my theory, also contributed to this outcome. Health policy became more complex in the nineties, due both to the emergence of health economics and the intricacy of the new Colombian health care system. Because politicians were unable to fully understand these more sophisticated policies, it became more risky to remove the technocrats who had designed them and who understand the system.

The chapter proceeds as follows. In the first section I use my theory of technocratic autonomy to explain technocrats' failure to entrench at the end of the NF. Next, I focus on the second health episode, providing evidence of experts' autonomy and their continuity in the Ministry of Health after the reform. In the third section, I detail experts' strategies to advance their goals in this intricate policy area and use my theory to explain experts' considerable stability. I conclude with some lessons derived from these two contrasting episodes.

HEALTH EXPERTS DURING THE NATIONAL FRONT

As discussed in Chapter 4, one of the NF's main goals was reducing poverty in Colombia (Hartlyn 1988; Dix 1967, 147). International reports produced before the NF period had pointed to health problems and the weakness of the health care system as direct causes of the country's underdevelopment (Sandilands 1990, 159–177; Ministerio de Salud, DNP 1990, 120–139; Abel 1996). The resource-starved health care system was also patronage-based: resources, facilities, personnel, and medicines were distributed according to political loyalty rather than to need (interview with Urrutia 2008). At the time, the *Instituto de Seguridad Social* (ISS) and other specialized insurance programs focused on urban, formally employed workers, while the Ministry of Health was responsible for the rest of the population. The NF leaders aimed to professionalize, decentralize, and depoliticize the health care system (Lleras Restrepo 1965; interview with Otto Morales Benites 2006).

During these years, the government paid considerable attention to health policy planning (Bradbury 1975, 6). The Ministry of Health was staffed with teams of medical planners and epidemiologists. As in other Latin American countries, these officials were mostly doctors, but in Colombia, they were also competent in public health and proposed a comprehensive health care system (USAID 1974, 18; Ministerio de Salud, DNP 1990, 148). These doctors included renowned experts in public health such as Luis Carlos Ochoa, Héctor Abad Gómez, David Bersh, Santiago Renjifo, and Francisco Yepes. All of them were part of international epistemic communities with connections to the United States, Chile, and Venezuela. These experts embraced modern doctrines promoted by the PAHO and other institutions that emphasized preventive measures to fight disease, rejecting dominant paradigms in Latin America at the time that focused exclusively on treating illnesses (Ministerio de Salud, DNP 1990, 139–144; interview with Ordóñez 2006).

NF governments increased health care expenditures in the country considerably. Although precise statistics about health care expenditures are difficult to find, a Fedesarrollo study shows that the ministry's per capita health care expenses increased by more than 250 percent over the NF period from 1966 to 1974 (Vivas et al. 1988, 286). Another study carried out by USAID shows a similar trend: health care spending rose from 1.8 percent of GDP in 1961 to roughly 3 percent in 1970 (USAID 1974, 149).

A variety of ambitious reforms accompanied these spending increases. In 1960, the Ministry successfully lobbied Congress to approve two legislative decrees (*Decretos* 1423 and 322 of 1960) that together established comprehensive health policies for the country (USAID 1974). Later, the second NF government (1962–1966) focused on health care in the countryside, training health promoters in rural communities in preventive medicine (Ministerio de Salud, DNP 1990, 150). Furthermore, various master plans aimed to provide guidelines for distributing resources and developing health care facilities (USAID 1974, 22–23). These plans focused on reducing political bargaining and electoral benefits as the criteria for resource allocation. A central goal of these reforms was to staff the health care system with a competent technical bureaucracy insulated from electoral pressures (interview with Ordóñez 2006). The government subsidized public health programs in national universities in an effort to raise the number and quality of health care professionals (Ordoñez 1968, 16).

These reform efforts peaked during the government of Carlos Lleras (1966–1970). Lleras appointed Antonio Ordóñez, an internationally recognized health expert and former PAHO official, as minister of health. Ordóñez brought with him a group of highly qualified experts to implement his ambitious reform plans (Ordóñez 1991; interviews with Ordóñez 2006 and Otto Morales Benites 2006). The team designed a national health plan (*Plan Nacional de Salud* 1968–1977) based on five priorities: developing health policy, evaluating and supervising decentralized health facilities, training personnel, encouraging more domestic health research, and creating a technical, non-politicized budget for the sector. These reform efforts were helped by a constitutional amendment in 1968 that gave the Ministry more regulatory power over the health system (DNP 1974, 31–33).

Ordóñez and his team reorganized both the administrative and regional apparatuses of the health system (DNP 1974, 23; Ordóñez 1991, 47). These reforms sought, among other goals, to relocate medical personnel to rural areas. Following his predecessors, Ordóñez also designed a plan to build health care facilities according to regional needs (*Plan Hospitalario. Construcción, dotación, y operación de hospitales para el trienio* 1970–1972, Law 39 of 1969). In addition, the Ministry launched vaccination and family planning campaigns while promoting the development of sewage systems (DNP 1974, 23). National investment in sewage infrastructure grew by more than 300 percent from 1965 to 1970 (DNP 1974, 77; interview with Ordóñez 2006).

These technical trends reversed after the end of the NF. Although the first post-NF government, led by conservative Misael Pastrana (1970–1974), continued these reforms, renewed electoral competition between liberals and conservatives in 1974 led health policy and the health bureaucracy to once again be used for clientelism (Ordóñez 1991, 47; Interview with Ordóñez 2006).

Paradoxically, this political intervention was partly brought about by the *técnicos'* relative success. Because the health care system was underfinanced before the reform episode, politicians saw few benefits from intervening in the health sector. Certainly, the health care system was used for clientelism, but it did not represent a substantial pool of resources. Because of the reforms during the NF, the health care system was better funded and had a greater national reach (Ordóñez 1991, 47; interview with Ordóñez 2006). Political competition also pushed politicians to reassert control over health policy. In 1970, former dictator Gustavo Rojas Pinilla from the opposition party ANAPO almost won the presidency. Facing this political threat, the incumbent had a strong incentive to directly manage health sector funds and social programs that could be used clientelistically in order to undermine support, which was especially high among the urban poor, for ANAPO. As a result, the nascent health technocracy declined.

What made health policy experts less resilient to political intrusion than were economic technocrats in the NPD or MF during those same years? The two factors related to the politics of policy in my theory that discourage politicization score low in this episode and explain much of this outcome.

First, as discussed in Chapter three, the political costs of poor health policy performance were low but, with the return of inter-party competition, potential gains from political intervention were considerable. According to Minister Ordóñez, reforms made the health care system a lucrative source of clientelism across regions and municipalities (interview with Ordóñez 2006). Politicians realized that they could use the system to their advantage with few consequences because the politically influential middle and upper classes were largely immune to the effects of intervention. These higher income groups were covered by the ISS, special insurance systems (*Cajas de Compensación Familiar*), or private insurers (USAID 1974, 19–20). Turning health care into a patronage-based system was attractive for politicians and incurred little political cost.

Second, no balanced constellation of stakeholders existed within the sector to protect the experts' autonomy. Liberals and conservatives did not check one anothers' control in this policy area, as they did in economic policy, because both enjoyed the spoils of clientelism at all levels of the system. The users of the public health care network, mostly the poor, lacked leverage while, as mentioned, other sectors of society had minimal interest in health policy. On top of this domestic imbalance, international financial and aid agencies chose largely to ignore this policy area. A 1974 study concluded that Colombia received little foreign aid for health care. Although organizations like PAHO, UNICEF, USAID, and a few private foundations were the most involved in Colombia (USAID 1974, 153), they did not possess the same leverage as IFIs did in economic policy. As a result, no stakeholders kept in check the power of political actors who exploited the system to their advantage. This exploitation, particularly at the regional level, robbed experts of initiative and was one of the main reasons the reforms did not continue after the sixties (Abel 1996, 140). In sum, the health sector was an attractive, unguarded supply of clientelistic resources that politicians gladly captured, leaving little space for technical autonomy.

Furthermore, the two factors related to the technical aspects of health policy were also minimally present. With respect to technical complexity, health policy at that time focused on distributing resources, which politicians could easily understand and dispute. Creating new positions in regional hospitals and building health care facilities or other infrastructure works such as sewer systems were all distributable goods that national and local politicians perceived as politically useful. The NF governments, for example, authorized the building of a hospital for a small town (*El Libano* in the Tolima region) where there was no technical reason for a facility of that size. Yet motivated by electoral interests, politicians insisted the investment was important to the development of the town and region (interview with Ordóñez 2006).

In addition, there was little technical consensus on health policy. As in other Latin American countries, divisions among health experts, especially between those prioritizing prevention and others focusing on treating disease, made reaching an agreement about adequate policies difficult (Ministerio de Salud, DNP 1990, 154). The result was that the experts' prescriptions were contested, meaning that politicians could appoint officials more in line with their clientelistic preferences.

Despite the intent of early reformists, the system did not experience drastic changes in its scope, coverage, or capacity. More importantly for this book, the Ministry remained technically weak. In the following years, Colombia's health ministry resembled others in the region: poorly funded, guided by prestigious doctors linked to political parties but largely without health policy credentials, and overburdened by its regulatory and health provision duties. By 1991, still only 20 percent of the population was covered by some form of social security, mostly the middle and upper classes by ISS (Ramírez 2004, 128). The rest of the population depended on public facilities managed by the Ministry of Health, which were quite difficult for the poorest citizens to reach.

LEY 100 AND "TEAM CHANGE" (1993–2013)

As in the previous episodes, experts arrived at the Ministry of Health in 1993 to confront a challenge that politicians were unable to resolve. President César Gaviria (1990–1994) appointed a team of health policy technocrats, acting as advisers to the Minister of Health, to push an ambitious health policy reform through Congress. This reform was one of, if not the most, comprehensive transformations of a national health care system in the history of Latin America (Nelson 2004, 49; Ramírez 2004; González-Rossetti and Ramírez 2000).[1] The adoption and ensuing implementation of this reform plan are clear evidence of experts' autonomy. Although they lost some leverage after Gaviria's administration, *técnicos* have remained relevant actors in health policy and, more recently, have regained saliency under a succession of three highly technical health ministers. In this section, I document these outcomes as well as provide evidence of experts' autonomy. In the next section, I use my theory to explain how technocrats achieved autonomy.

The 1993 health policy reform occurred as part of a wider set of political and social reforms adopted in the 1991 Constitution. A growing crisis fueled by corruption, internal warfare, and drug violence led to a national debate about how to strengthen democracy in Colombia. President Gaviria embraced the demands for reform and supported the election of a Constituent Assembly in 1990. For various reasons, reformists from traditional parties and new political organizations, such as the recently demobilized leftist guerrilla group M-19, ended up dominating the assembly (Falleti 2010, 135–139). Some of these reformist assembly members pushed to include ambitious social rights in the constitution to re-legitimize the political system (Ramírez 2004: 126–127; Falleti 2010, 138–139). At the same time, Gaviria's government and other assembly members sought to institutionalize market reforms in the Constitution. These included so-called second generation reforms, such as overhauling social policy (health and education) and the justice system (Edwards and Steiner 2008; Cepeda Ulloa 1994), that IFIs had recently begun encouraging.

In this way, two different reform agendas overlapped in the assembly: a social-progressive project and a neoliberal one. These two positions emerged in the debates about the constitutional articles regarding health policy. On one hand, some representatives of the ISS, medical associations, and health care workers' unions demanded a single health care system led by the state and based on equity. On the other hand, members of the government proposed a

[1] For a comprehensive analysis of the debates, adoption, and implementation of the reform, see González-Rossetti and Ramírez 2000. Also see the Health Ministry's document "Recuento del Diseño y Trámite de la Reforma en Seguridad Social en Salud" (Ministerio de Salud 1994).

system that promoted efficiency and market competition with more private sector participation (Ramírez 2004, 128–129; Uribe 2004, 206–207).

The final articles (48 and 49 of the 1991 Constitution) were a compromise between expanding health care coverage through the state and efficiency by involving market forces. The Constitution mandated that, in a year, the government had to present bills to Congress detailing the content of these constitutional articles. Before adjourning, the assembly established a commission, including social and state actors involved in the health care system, with far-reaching powers to debate social policy laws, including health.

The members of the commission remained divided between these two camps. NPD technocrats representing the government, proposed a law that promoted private participation in all areas, including insurance, as well as other poverty-reduction policies following recent IFI proposals. On the other hand, medical associations and experts from the ISS who embraced a primary health care approach proposed state-centered reform, which limited private participation in health care provision (Ramírez 2004, 129; González-Rossetti and Ramírez 2000, 37–43). These actors denounced the NPD's proposal as abandoning the equity-enhancing and social content of the 1991 Constitution (Ramírez 2004, 129; interview with Giedion 2008). To make matters more complicated, the health ministry took a third position. Shortly after the assembly, Gaviria had invited the M-19 into the government coalition to lead reform in the Ministry of Health. Although the Ministry did not officially take sides in the debate, its positions were closer to the ISS than to the NPD (interview with Paredes 2008).

As a result, the government never presented a unified health policy reform proposal, and debate stagnated. Partly due to these impasses, M-19 left the government in November of 1992. Gaviria took the opportunity to change the leadership of the Ministry of Labor, Ministry of Health, and ISS to achieve a unified reform team capable of pushing forward social security reforms. Gaviria appointed Juan Luis Londoño, who had a PhD in development economics with intensive training in health policy economics, as his health minister. Prior to this appointment, Londoño was a deputy director at the NPD, where he had advanced the aforementioned proposal.

Once in the ministry, Londoño appointed a group of *técnicos* as his advisers that the press dubbed "Team Change" (*Equipo Cambio*). This team was recruited from NPD officials working on social policy and others who had recently received postgraduate degrees in public health (González-Rossetti and Ramírez 2000; interview with Paredes 2008). With ups and downs, this reform team eventually acquired a considerable presence, and then autonomy, within the Ministry.

To push through reforms, Londoño and his team undertook a broad, participatory process to determine the content of the new law. This process included consultations with government officials, regional authorities, the Colombian Medical Federation, private health institutions, and other groups

(Ramírez 2004, 132; interview with Paredes 2008). The Ministry took these consultations seriously, amending its original stance to include equity considerations and eventually producing a proposal that gained acceptance among more sectors than did previous reform efforts. The draft consisted of broad, general principles, leaving the details for the implementation stage (González and Ramírez 2000, 50–51). With the crucial backing of President Gaviria, *Ley 100* (Law 100) was finally approved in December of 1993. The government tied health reform to the overhaul of social security and was thus able to pass both simultaneously (González-Rossetti and Ramírez 2000).

The resulting *Ley 100* places all citizens into a single health care system but with two tiers or *regimes* based on ability to pay.[2] The contribution-based regime, including the formally employed sector and independent workers with sufficient income, allows beneficiaries to choose from health plans (*Plan Obligatorio de Salud*, POS) offered by private insurance companies called Health Promoting Enterprises (*Empresas Promotoras de Salud*, EPS). The state collects a payroll tax from workers to then pay EPSs, which in turn contract service providers, both private and public, which are collectively called Health Service Providers (*Instituciones Prestadoras de Servicios*, IPS). Colombians without sufficient income, as determined by the DNP, are covered under the subsidized regime, which has the same structure as the contribution-based regime. Part of the fee paid by those in the contribution-based regime funds the subsidized system while the government supplied the difference (Escobar et al. 2009, 5). The two regimes offer different levels of coverage, with the subsidized system focusing on the basic services recommended by selective health care strategies and the contribution-based regime providing a more comprehensive package.

The system attempts to address both the equity and efficiency concerns raised during the debate surrounding the 1991 Constitution. Although it began with two unequal tiers, *Ley 100* expanded basic health care coverage and mandated that, by 2001, both systems converge so that all citizens would enjoy the benefits of the contribution-based regime. At the same time, by working through private insurers, as well as a mix of public and private service providers, the law sought to encourage competition that would produce efficiency gains over the old system of providing supply-side subsidies to public hospitals (Escobar et al. 2009, 4–6).

If approving the reform demonstrated experts' leverage and saliency, the events after the reform solidified this impression. Experts successfully implemented many reforms quickly and under stressful conditions. Several factors made the already daunting task of transforming the health care system more difficult. First, the reforms required harmonization with political decentralization that the new constitution also initiated. The Constitution and some laws,

[2] For a more detailed explanation of this complex system see Ramírez 2004, 134–136; González-Rossetti and Ramírez 2000, 32–35; Uribe 2004, 197–201.

both old and new (e.g., *Ley 10* of 1990), transferred a myriad of responsibilities to regions and municipalities, among them regulation of healthcare, financing, and provision. Experts faced the challenge of coordinating subnational governments to carry out these functions, which were critical to the success of the new system, especially to its subsidized component.

Londoño and his team also worried that whoever won the 1994 presidential elections would either reverse Ley 100 or halt its implementation. This concern was well grounded. The candidate of the incumbent liberals and favorite to win the election, Ernesto Samper, came from the party's social democratic wing that was often in conflict with Gaviria's more neoliberal faction. Samper clearly articulated his opposition to expanding the private sector's role in health care. When he became the liberal party's presidential nominee, the team knew they had limited time to implement the reforms (interview with Paredes 2008).

Experts abandoned their earlier consultative strategy for a closed, secretive process to implement the law. Team Change worked frantically over six months to entrench the reform. Although experts attempted to vet these policies through a new National Council for Social Security in Health (*Consejo Nacional de Seguridad Social en Salud*), an institution that included representatives from all parties involved in health policy and was responsible for regulating the system according to *Ley 100*, this body lacked the capacity to effectively supervise experts. Londoño and his team passed twenty-five decrees in a mere seven months (González-Rossetti and Ramírez 2000, 50–53; Ramírez 2004; interviews with Giedion 2008 and Paredes 2008). Reform opponents argued that Team Change selectively read the law, making the system much more privately oriented than originally intended, while members of the team claimed that they were simply following *Ley 100* (González-Rossetti and Ramírez 2000, 50–51).

Knowing that these regulations were easily reversed, especially because of the general language of *Ley 100*, Team Change prioritized creating the new institutions included in the law. Experts believed that these institutions would generate new stakeholders that could oppose a reversal in the system (González-Rossetti and Ramírez 2000, 50–53). These stakeholders included the agencies in charge of purchasing health care services, EPSs in the contribution-based regime, and *Administradoras de Régimen Subsidiado* in the subsidized regime. Health policy technocrats also adopted new regulations for health care providers, both private and public.

Experts also aimed to maintain technical continuity in the ministry after the change of government. Knowing that the complexity of the reform exceeded the capacities of the ministerial bureaucracy, Londoño tried to build technical groups that could work with some independence inside the ministry. Starting in 1994, a team of experts dubbed "the Harvard Group," named for the consulting support they received from the Harvard School of Public Health, was charged with designing a plan to provide technical support to reforms in the following years. This group produced what is known as "The Harvard Report"

(Colombia Health Sector Reform Project 1996), which included a blueprint for implementing the reform.

The Harvard Report proposed using a foreign loan to hire an internationally reputed academic institution to provide technical assistance and oversee implementation (interviews with Castaño 2008 and Ramírez 2009). This plan was never realized due to opposition within the Ministry of Finance, which found such a loan too costly. Nonetheless, government funds and an IDB loan for $65 million (U.S. million) allowed for the creation of a special office, *Programa de Apoyo a la Reforma en Salud* (PARS), to support the implementation process. Starting in 1996, PARS took over some of the responsibilities identified in the Harvard Report with Nelcy Paredes, Londoño's right hand during the initial reform period, as its first director.[3]

When Samper won the 1994 election, he appointed a doctor and well-known reform opponent as Minister of Health. This minister brought his own team of experts, who came largely from the ISS and had opposed the NPD's proposals during the early reform debates (Ramírez 2004, 139, Uribe 2004, 208). The new minister delayed implementing further reforms and reversed some regulations, especially those concerning demand-side subsidies for public hospitals. The new regulations restored abundant subsidies for public health care providers and forced EPSs in the subsidized system to contract a percentage of services from these facilities. These changes were at odds with the original design that had aimed to promote efficiency by making public facilities compete with private service providers (interview with Ramírez 2009). However, the minister left after one year and Samper's ensuing ministers, as well as those of his successor conservative Andrés Pastrana (1998–2002), were generally more sympathetic to the reform agenda (interviews with Galán Sarmiento 2009 and Paredes 2008). Nonetheless, provider subsidies remained in place.

In 2002, President Álvaro Uribe reappointed Juan Luis Londoño to invigorate the reform process, especially with regard to expanding health care coverage. Reform had encountered fiscal problems due to the 1999 international financial crisis, judicial decisions that required the state to purchase medicines and medical services not included in the original POS, and difficulties brought on by health decentralization (Uribe 2004, 104; interviews with Uribe 2008, Santa María 2008 and Pinto 2009). Londoño brought back several members of Team Change as well as new healthcare specialists. He also proposed merging the labor and health ministries into a single "Ministry of Social Protection," which Congress approved that year.

When Londoño died in a plane crash shortly after taking office, Uribe appointed Diego Palacio, a member of Team Change but not a credentialed expert, to head the ministry. Palacio retained several members of Londoño's team and remained in office until the end of Uribe's second term in 2010

[3] PARS reports and publications are found in minproteccionsocial.gov.co/salud/Paginas/Programa%20PARS.aspx

(interview with Pinto 2009). Palacio focused on expanding health care coverage, which grew astronomically from around 60 percent in 2002 to 90 percent in 2010, but paid less attention to controlling the system's growing deficit (interviews with Zapata 2011, Ramírez 2009, Giedion 2008, and A. Gaviria 2008). According to information from the Ministry of Health, 92 percent of Colombians now enjoy some form of health care coverage within the system: 49 percent in the subsidized regime, and 43 percent in the contributive regime. The rest are either in special systems or not affiliated. Additionally, in response to the growing complexity of the system, the ministry created the position of technical vice minister (interview with Ramírez 2009).

The end of Uribe's administration was marked by a severe financial crisis in the health sector that resulted from the aforementioned problems but also from new rulings by the Constitutional Court that placed additional stress on the system (Clavijo and Peña 2010; Fedesarrollo 2009; 2010). The government declared the system to be in a state of emergency and issued executive decrees in response, although the Constitutional Court later reversed them (Clavijo and Peña 2010). This crisis eventually led to an increase in technocratic autonomy in the ministry.

Because of these growing problems, health policy gained public salience. Several critical voices even demanded a reversal of the 1993 reform. President Juan Manuel Santos (2010–), Uribe's successor, appointed Mauricio Santa María, a widely recognized expert who was a deputy director at the NPD and then at Fedesarrollo, as Minister of Social Protection to tackle the system's serious fiscal problems. Santa María had long worked on the financial problems of social policies in Colombia and warned of the disastrous consequences of growing deficits. The new minister appointed qualified experts to address the immediate crisis and establish a sound, long-term fiscal base for the system (interviews with Zapata 2011 and Reyes Forero 2012). To confront these financial problems, the team enacted major reforms, which included decoupling the Ministry of Labor and the Ministry of Health.

In December 2011, Santos announced that Santa María would become NPD director and Beatriz Londoño, Santa María's technical vice minister and a highly reputed expert, Minister of Health. These moves guaranteed high technical continuity in the ministry. More recently, in August 2012, Alejandro Gaviria, a widely respected social policy expert who was Dean of Economics of *Los Andes* and one of the harshest critics of the financial crisis in the health sector, became Santos's third Minister of Health. All these ministers have pushed for reforms, some quite controversial and drastic, to address the system's serious budget problems.[4] As I write, Colombia's health care system remains an important political issue.

[4] *El Tiempo*, "La Gente Ya No Tendrá que Afiliarse a una EPS: Alejandro Gaviria," February 25th, 2013; *El Tiempo*, "Gobierno Acepta Ajustes, pero Defiende Continuidad de Reforma de Salud," November 5th, 2013.

In conclusion, health policy technocrats were more successful than their 1960s counterparts at achieving considerable autonomy and continuity within the Ministry of Health. After Londoño left the government in 1994, experts remained relevant actors either as advisers in the Ministry's office, in special offices crated to give technical support to the ministry (PARS), and more recently again in the top positions of the Ministry such as the technical vice ministry. However, these successes have accompanied failures that reveal the limits of experts' autonomy. PARS, for example, provided a safe place from which to maintain some technical autonomy in the ministry, but it lacked control over the reform process and could not directly implement its own recommendations (interviews with Ramírez 2009 and Pinto 2009). In addition, even if technical appointments to head state institutions are now more common, few would conclude that this is an entrenched pattern in Colombia. Health policy experts in this period enjoyed more autonomy than during the NF or at any point in time in Peru, but less independence than did the economic experts discussed in previous chapters. What explains this intermediate outcome?

EXPLAINING TECHNICAL AUTONOMY AND CONTINUITY IN AN UNLIKELY POLICY AREA

In this section, I will first discuss how experts' actions helped build a moderate amount of autonomy and then how medium scores in three factors of my theory explain this outcome in what constitutes a "less-likely case" for technical independence. These three factors are (1) elevated political costs of poor policy performance, (2) a somewhat balanced constellation of stakeholders, and (3) a rise in health policy complexity. The first two "politics of policy" factors increased politicians' need to appoint experts and limited intrusion into health policy. The last factor, related to the technical aspects of policy, contributed to limiting political intrusion by making health policy less comprehensible for politicians, thereby complicating politicians' assessment of the potential costs of intervention. The remaining factor, technical consensus, remained low. Although recent decades have seen a global consensus emerge around health policy, debates about what constitutes adequate health policy in Colombia have continued since the early nineties.

As in the previous two chapters, this episode shows how expertise provided technocrats with considerable leverage over other actors. In a reform process that became increasingly technical, experts gained control over the policy content of the reform. Londoño and his team were firmly backed by a reformist president, providing them with an advantage not enjoyed by experts in some of the other episodes I discuss. Yet their actions and strategies were also important to achieving the reform and guaranteeing technical continuity (González-Rossetti and Ramírez 2000, 66–79). This episode illustrates three types of strategies through which experts built their own autonomy.

The first of these strategies was purposively using their expertise, particularly of health economics, to exclude opponents from the policymaking process and overcome resistance from less well-trained individuals. As explained by a member of Team Change interviewed by González-Rossetti and Ramírez:

> I was asked to present the proposals to interest groups and congressmen who came to discuss the proposals. ... It was a strategy that facilitated the international support of the process, as well. Finally, it was also a strategy for exclusion. People who did not understand could not argue the main aspects of the reform. ... And Juan Luis Londoño did not face many technical arguments against the reform: probably the role played by Fedesarrollo was one of the few important ones in that sense. Londoño only feared Fedesarrollo's arguments, he wanted to be sure the data were right when they had to discuss [the data] with him.(González-Rossetti and Ramírez 2000, 75)

As discussed later, when analyzing the policy complexity factor identified in my theory, the reform's multifaceted regulations and its focus on economic aspects of the health system, such as efficient resource distribution, made this strategy more effective.

A second strategy was to establish alliances with domestic and foreign actors in order to advance the reform (Ramírez 2004, 131). Aware of IFI interest in health policy reforms, experts sought these institutions' support for both reform and technical continuity in the ministry. The Harvard School of Public Health provided abundant technical advice during the reform process, bringing to Colombia some "of the best people working on health topics in the world" (González-Rossetti and Ramírez 2000, 75). At the domestic level, the team built horizontal and vertical alliances with other governmental agencies working on reforms, such as the NPD, the MF, and technical teams in the ISS and the Ministry of Labor (González-Rossetti and Ramírez 2000, 66–72). These technical networks within the Colombian state facilitated coordination to pass the reforms.

Lastly, experts sought to entrench expertise in health policy by creating semiautonomous offices within the ministry. The Harvard Team (1994–1996) and then PARS (1996–2008) worked in the ministry but were not part of the ministerial bureaucracy. These offices remained quite autonomous, independently receiving and managing external funds. Harvard's School of Public Health also contributed to experts' continuity, as members of the team carried out studies at Harvard during or after Londoño's first term. As discussed later, experts also inadvertently contributed to their own stability by rushing to create new institutions within the health care system. Although their goal was to entrench the reform, these institutions ended up increasing the need for expertise in the sector.

Nonetheless, experts' strategies alone do not explain this intermediate outcome. As discussed in the next chapter, Peruvian experts were also quite strategic but failed to achieve similar continuity. The conditions in this episode were unusually favorable for experts working in health policy.

Politics of Policy Factors

In the first health episode, NF politicians were able to exploit the health care system because it was not of immediate concern for many Colombians, especially in the more vocal middle and upper classes. After the 1993 reform, the cost of poor policy performance became considerably higher and solving health policy problems became much more urgent for the incumbent. This higher cost came from expanding coverage of the public health care system. More important for increasing this political cost than just numbers is the fact that middle and upper class interests in the contribution-based system, which served 43 percent of the population in 2012, are now entangled with poor Colombians in the subsidized system. Although the two regimes are largely separate (only a small percentage of the contribution-based fund is shared with the subsidized regime, for example), any crisis affecting the Ministry of Health puts both groups at risk.

However, the costs of poor policy performance were not immediately evident for politicians after the reforms were passed and only became apparent as problems surfaced. As described earlier, several issues, largely related to poor allocation of resources and a growing deficit, plagued the reform following its adoption. In these first years, politicians again profited from a better funded health system. A significant percentage of resources were transferred to regions and municipalities where they were frequently used for clientelistic ends, such as creating additional patronage positions in public facilities, or simply lost to corruption (interviews with Santa María 2008 and Paredes 2008). From an average of 1.2 percent of GDP in the eighties, the system grew to around 3.6 percent of GDP in 1998, of which local governments spent around 70 percent.

But clientelism and corruption were not the only problems. Maintaining subsidies for public facilities led to an increase in health care expenditures, since the government was now subsidizing both health care providers and consumers because of the failure to enforce competition. Additionally, the number of medical procedures authorized in the contribution-based and subsidized systems grew as citizens used administrative or judicial rulings to demand that the government pay for more procedures (Escobar et al. 2009; Glassman et al. 2009; Ramírez 2004, 136–137; interviews with Cepeda 2006, Santa María 2008 and Zapata 2011). In 2008, new Constitutional Court decisions put the system under even more stress by significantly expanding health care expenditures.

As a result of these processes, the health care system nearly suffered financial collapse with important repercussions among the public (interview with Reyes Forero 2012). President Uribe faced strong public criticism for what was perceived as inaction regarding the health care system. Opponents of the 1993 reform, including health experts, reemerged and blamed the crisis on private health organizations, especially EPSs. Uribe approved an emergency decree to cut some of this deficit, but the Constitutional Court declared it unconstitutional. All these events put the health care crisis at the center of public debate.

It is difficult to measure this growing public concern. One indicator is the considerable attention that health policy gained in the press. A search for the terms "salud" (health) and "POS" (*Programa Obligatorio de Salud*) in *El Tiempo*, Colombia's most widely read newspaper, yielded fewer than 15,000 results in the decade before 2006. The number of results for the same search in 2007 jumped to 19,394, then 30,549 in 2008, 30,626 in 2009, and 28,693 in 2010 (an election year).[5] Additionally, in clear contrast to the recent past when security issues were the dominant concern of Colombians, social issues took center stage in the 2010 presidential election. According to a Gallup poll from April 2010, health care ranked second with 12 percent of respondents identifying it as the most acute problem in Colombia, after unemployment (39 percent).[6] As mentioned, Juan Manuel Santos, Uribe's successor, responded to these concerns by appointing experts respected for their knowledge of these financial problems to be his health ministers. Experts serve as public symbols of politicians' commitment to addressing public concerns.

This growing cost of poor policy performance and the potential consequences of the system collapsing opened more space for expert autonomy and increased technocrats' leverage in the health sector. It was now clear to politicians that there was less tolerance for poor policy performance. Colombia's health care system is still in crisis and is currently one of the most highly contested issues in the country.[7] This political cost becomes more daunting for politicians when paired with the growing complexity of health policy. The intricacy of the health care system raises the technical level needed to understand the effects of policy decisions, giving experts more space to exercise their autonomy. In conclusion, this higher political cost and the potential consequences of a full-blown crisis boost expert autonomy.

Second, a more balanced constellation of stakeholders emerged in health policy, providing experts with more space to act autonomously. This constellation was first enhanced by the growing interest of IFIs in health policy reform during the nineties. As mentioned, members of Team Change were part of international epistemic communities, because many of them were working or had worked for IFIs (González-Rossetti and Ramírez 2000, 60–62). The neoliberal reforms advanced by Gaviria included issues such as health policy and judicial reform about which IFIs had recently begun advising, as part of what they called "second generation reforms." Politicians were aware of this

[5] *El Tiempo*, archivo electrónico. http://www.eltiempo.com/seccion_archivo/index.php

[6] *Semana*, "¿Cuáles Son los Temas que Llaman a Votar en Estas Elecciones?" Online edition. April 29th, 2010. http://www.semana.com/politica/cuales-temas-llamaran-votar-estas-elecciones/138254-3.aspx

[7] *El Tiempo*, "Pese a Algunos Avances, la Salud en Colombia Todavía Sigue en Crisis," January 23rd, 2012.

growing interest, reinforcing the need to appoint experts in these areas who could act as interlocutors of IFIs.

Although IFIs had limited involvement in Colombian health policy prior to 1993 (Uribe 2004, 108), the reform process increased their presence in this policy area. The WB and the IDB supported the reform in many ways, such as by providing funds to offer competitive salaries to some members of the technical team working on special projects during the reform period. Additionally, as mentioned, another foreign institution, the Harvard School of Public Health, provided support from 1993 to 1996 (Ramírez 2004, 138). By becoming involved in reforms, IFIs created a more balanced constellation of interests around health policy and thereby increased expert autonomy.

As discussed in the next chapter, IFI interest in health policy reform had diminished by the end of the nineties. While the withdrawal of international agencies in Peru tipped the balance of stakeholders away from reform, by the time this happened in Colombia, other domestic actors had emerged that were powerful enough to counterbalance other influential interests. Insurers and providers now had a stake in keeping experts in the system in order to limit political appetites and curtail other stakeholders' influence. These actors exerted themselves through two associations that defended private interests in the system. The Colombian Association of Health Purchaser Companies (*Asociación Colombiana de Empresas de Medicina Integral*, ACEMI) lobbied for all EPSs in the public and private systems,[8] while the Association of Health Providers (*Asociación Colombiana de Hospitales y Clínicas*, ACHC) included both private and public facilities.

Private health insurers and providers now have stronger interests in public health policy that makes them exercise checks on other actors, precisely as the system's designers intended (Londoño and Frenk 1997). ACEMI, for example, produces frequent opinions about health policies.[9] Similarly, the ACHC exerts its influence on the government and other actors of the system, such as health purchasers.[10] Given the costs of the current system for the private sector, even the powerful Association of Financial Institutions (*Asociación Nacional de Instituciones Financieras*, ANIF) voices frequent technical opinions about the costs and limitations of the health care system (ANIF 2008; 2009; 2010). These mutual checks and balances allow experts to exercise their autonomy.

Critics of the reform highlight the power of these private actors to shape policy in their favor. The fact that some health experts involved in the reform now work in these associations only reinforces the image of a cozy relationship

[8] See, for example, the frequent press releases produced in response to policies during the current health crisis: http://www.acemi.org.co/index.php/es/comunicados

[9] See, for example, the frequent press releases produced in response to policies during the current health crisis: http://www.acemi.org.co/index.php/es/comunicados

[10] For an example of technical reports and research produced by ACHC, as well as press releases, see: http://www.achc.org.co/investigaciones.php?idcat=28

between experts and private institutions. Yet overall, these relations do not signify that experts are subordinate to the private sector. Experts' preferences frequently conflict with the interests of private actors. Especially in recent years, experts have sought to put tighter controls on these institutions, specifically health insurance. Currently, experts are pursuing more oversight of EPSs because of allegations of corruption against them over the last several years. These actions include financial sanctions and interventions in some of these institutions, including some of the most powerful EPSs of the country.[11] Similarly, experts have sought closer control over these private actors. The regulatory agency that had originally intended to monitor these new private actors, the *Superintendencia de Salud*, did not function as expected (interview with Uribe 2008 and Reyes Forero 2012). Recently, this agency's responsibility to oversee health insurance was transferred to the powerful *Superintendencia Financiera*. Experts argued that the *Superintendencia Financiera*, which is a strong and respected government agency that effectively controls financial institutions, is better suited for this task.

Further, the idea that unified private interests have captured public policy does not acknowledge the divisions that exist between these private interests. There are several ongoing disagreements between insurers and providers, especially about the immense debts EPSs (insurers) owe to IPSs (health care providers) that are partially responsible for the system's precarious financial situation. Both sides have proposed solutions that the other strongly criticizes, and insurers have even supported ministerial actions against providers. Thus private actors in the health sector are at odds with one another and contribute to a more balanced constellation of stakeholders that, in turn, increases experts' autonomy.

This constellation is still not as balanced and plural as in economic policy. Some crucial actors such as associations of patients and civic groups are still relatively weak (Ramírez 2004, 132). Also, as mentioned earlier, some of the new institutions, such as the *Superintendencia de Salud*, have not functioned as expected, thereby limiting pluralism. Similarly, the Health Council that was supposed to oversee the system never gained institutional strength (Grupo de Economía de la Salud 2007; Naranjo Velandia et al. 2002). The Council was replaced in 2007 by a health commission with similar functions (Comisión de Regulación en Salud, CRES), which itself has since been dissolved. Even so, the reforms helped create powerful stakeholders that led to a more favorable environment for expert autonomy.

Technical Aspects of Policy Factors

An increase in the complexity of health policy, one of the factors related to the technical aspects of policy, also favored greater technical autonomy. As

[11] *El Tiempo*, "¿Por qué Renunció el 'Defensor' de las EPS?" June 3rd, 2011.

discussed in Chapter 3, an increasing focus on economics in the late eighties and early nineties made health policy more complex. The health economics approach emerged in response to the challenges of meeting health care needs where resources are scarce. Health policy now required understanding econometric models and complex statistics. *Ley 100* included diverse technical and financial aspects that only individuals with proper training in economics could fully grasp. Team Change was amply trained in this discipline: "The group liked to talk based on numbers and graphs" (González-Rossetti and Ramírez 2000, fn 85).

Health policy complexity also grew because of the myriad institutions involved in the Colombian health care system (Giedion et al. 2000). Understanding the intricate network of new and old institutions and their relationships with each other is difficult even for experts accustomed to navigating the system. To make matters worse, the system became even more complex due to the decentralization reforms that multiplied the institutions involved in health policy. Colombia has approximately 1,000 municipal and 32 regional governments, which lacked technical expertise but were each responsible for health policy in their jurisdiction.

This complexity furthered politicians' dependence on experts. These technical intricacies make it very difficult for new ministers to fully comprehend their responsibilities and the potential problems the sector faces, which reinforce their reliance on experts as advisers. According to former minister Augusto Galán Sarmiento (1995–1996), health ministers need capable experts that can understand the system and respond to its complex challenges (interview with Galán Sarmiento 2009).

Technical complexity also gave experts the upper hand when debating health policy during and after the reforms. The majority of politicians, doctors, union members, and other health experts lacked similar technical knowledge. As discussed earlier, experts used this complexity to exclude nontechnical actors from the reform process. Similarly, Teresa Tono, PARS Director from 2003 to 2008, mentioned that opponents of the reform frequently approached the president or health minister to criticize the system. These criticisms were usually deflected by showing their lack of empirical support. While the Ministry and PARS had abundant technical studies backing their positions, many critics based their arguments on anecdotal evidence (interview with Tono 2008). Complexity, thus, contributed to experts' autonomy and continuity.

The fourth factor, expert consensus, however, remains low in Colombia. Although there is growing agreement among experts about the need to focus on resource allocation when debating health policy reform, the same divisions debated during the 1991 Constituent Assembly persist today. Many health policy experts support the reform, but an important group of specialists from public universities, doctors' associations, and public hospitals remain quite critical (Redondo and Guzmán 1999; Yepes et al. 2010). Reform opponents criticize new private institutions, mainly health insurance, for not allowing for real equity in the system and channeling disproportionate resources toward the contribution-based regime. These experts accuse the technocrats who created *Ley 100* of following in the footsteps of traditional Colombian health officials, with their narrow focus on curing illness, rather than on adopting a broader preventive perspective (Yepes et al. 2010).

These divisions among experts weaken technocrats' power relative to politicians' power because politicians can selectively appoint experts with views more closely aligned with their own interests. This happened when Samper's first minister appointed the group of ISS experts who opposed the reform. At the same time, politicians from the liberal party's left wing and the new leftist party, *Polo Democrático*, frequently used these experts' ideas to criticize the market legacies of Gaviria's administration. These divisions became more prevalent recently when the health sector crisis brought the whole reform under scrutiny. Some experts demanded an alternative, more state-centered approach that could overcome many limitations of the current system (Yepes et al. 2010). For now, supporters of maintaining the current health reforms remain in charge, but these disagreements are still strong. Although the current Minister, Alejandro Gaviria, favors the overall scheme of the reform, he is pushing for some deeper changes to the system in order to guarantee its financial stability.

Even so, it seems unlikely that politicians will drastically reduce the technical level involved in policy making: politicians need experts from one side or the other in the Ministry. The three other factors leave less space for politicians to significantly limit the role of expertise, as happened in Peru when IFIs lost interest in health policy and politicians clashed with experts. Even if new experts with stronger reform prescriptions enter government, they will almost certainly have similar technical credentials to those of current experts. Given the political environment in Colombia, the cost to the incumbent of not doing so is potentially quite high.

CONCLUSION

These two episodes show that experts can gain autonomy even in unlikely policy areas such as health. In both episodes, experts enjoyed considerable leverage to adopt their preferred reforms and a good deal of independence from politicians and other social actors. Health policy technocrats in the sixties built up a public health system and their counterparts in the nineties achieved ambitious reforms. Nonetheless, the low scores in all factors of my theory in the first episode led to the demise of experts when open political competition returned after the NF. This first episode highlights how vulnerable experts' autonomy can be within this social policy area.

On the other hand, higher scores in my two "politics of policy factors" as well as in technical complexity contributed to greater entrenchment in the second episode. Higher costs of poor policy performance, a more balanced constellation of stakeholders, and increased complexity all made experts integral to health policy, which forced politicians to keep experts in the state and give experts more autonomy. These higher scores have persisted for the last two decades and allowed experts to remain relevant in health policy during that time. In sum, these two episodes lend support to my theory of technical autonomy.

What else can we learn from these episodes? First, as discussed in my theory, politicians have incentives to intervene in the health sector and generally encounter low political costs in doing so. Both episodes show the danger of delivering more funds to state policy areas without considering this political ambition. Policy makers should be aware that in some sectors of the state, such as a decentralized health system, politicians' considerable influence opens the door for clientelism and corruption.

In addition, the second episode shows how reforms themselves can enhance the potential and real political costs of mismanaging a policy sector. By including middle and upper classes in the health care system, the reforms made health policy more relevant to a bigger and more vocal sector of the population. Exit strategies were scarce for these well-off sectors, which meant they had incentives to hold politicians accountable for poor health policy performance. As a result, experts became more important for politicians.

Similarly, this episode shows how policy reform itself can increase the number and importance of stakeholders within a policy sector, thereby indirectly making technocratic autonomy more likely. Experts rushed to create institutions to entrench the reform, which also fostered better conditions for their own survival. These increased constellation of stakeholders help to explain the persistence of technical autonomy even when IFIs lost some interest in health policy reform at the end of the nineties. The contrast with the ensuing Peruvian episode reinforces these conclusions.

7

Health Technocrats in Peru (1990–2002)

Health experts in Peru were appointed to the Ministry of Health (MINSA) during Alberto Fujimori's first administration (1990–1995). Fujimori's first minister of health hired experts to help him reform the health sector and attract external loans from IFIs interested in health policy reform (Vidal 1994a). Even though this minister was soon fired, some of the experts that he appointed remained at MINSA until 1993, when they gained considerable autonomy. Specifically, experts in the International Cooperation Office (*Oficina de Cooperación Externa,* OCE) designed and pursued reforms that significantly changed the health care system. Even if not as comprehensive as the Colombian reforms, the changes OCE spearheaded were quite notable for a sector characterized by immobility (Ewig 2004).

Despite traditionally high turnover rates for ministers, directors, and advisers in MINSA, experts in the OCE remained in their positions through seven ministerial terms during Fujimori's administration (1990–2000). Nonetheless, this technical group lacked the leverage to push forward more ambitious reforms, limit political intervention in some instances, and ultimately to entrench its autonomy in the Ministry of Health. After a series of political clashes and the reduction of external funds, technical influence gradually weakened and lost its relevance in 2001.

Even though it proved to be only temporary, what explains the autonomy and short-term stability of experts in this "less likely" policy area? What conditions limited this autonomy in the Peruvian episode and made technocrats vulnerable to dismissal when they clashed with politicians in the early 2000s? I argue that medium scores in two of my theory's factors were crucial for experts' autonomy. First, the involvement of IFIs in health policy reform during the nineties created a more balanced constellation of stakeholders in the health sector. This external demand pushed politicians to appoint experts and give them more autonomy. Second, the increasing prevalence of health economics made policy in this area more complex, thereby limiting political intrusion. Nonetheless, these favorable conditions did not suffice to permanently entrench technocratic autonomy, especially when IFIs gradually lost interest in health policy reform at the end of the decade. Although experts are now more present in MINSA than they were in the eighties, this institution remains an area of low technical autonomy.

This chapter proceeds as follows. In the first section, I provide some background on the Peruvian health care system and briefly examine a failed reform effort in the eighties that shows the challenges health reformers faced in Peru. In the second section, I discuss the rise of experts within MINSA, provide evidence of their significant autonomy, and detail the reasons behind their eventual dismissal. In the third section, the chapter's core, I use my theory of technocratic autonomy to explain both experts' autonomy and their demise. Particularly, I discuss how even though experts achieved significant stability in their positions, technocracy did not entrench and remained vulnerable to changes in political preferences. In this episode, reforms did not create stakeholders with interests in public health policy nor did they include middle and upper classes in the system. I conclude by highlighting how this episode contrasts with the more successful Colombian one.

THE HEALTH MINISTRY BEFORE THE REFORM

MINSA has always been a weak and underfinanced state institution, falling very short of meeting its vast responsibilities.[1] MINSA's main duty is to oversee the health care system, both public and private. However, MINSA also shares responsibility for public health provision with the social security system for the formally employed, formerly called the Peruvian Social Security Institute (*Instituto Peruano de Seguridad Social, IPSS),* and now, Health Social Security (*Seguro Social de Salud*, ESSALUD).[2] MINSA facilities provide health care to almost 75 percent of the population, mostly from the poorer sectors. Currently, 31.4 percent of Peruvians are covered by a subsidized public health insurance (*Seguro Integral de Salud* (SIS), 24.3 percent by social security, 2.8 percent by other special public systems, 1.5 percent by private providers, with around 40 percent still lacking insurance (ENAHO 2012; Chon et al. 2011; Petrera and Seinfeld 2007). Even though poverty is more pervasive in rural areas, the health system has traditionally been highly urban (Ewig 2010, 45–56, 108;

[1] For a historical overview of several problems in the Peruvian health sector, see Alvarado Pérez and Lenz 2006; Chon et al. 2011; Ewig 2004, 218–226; Francke 2001; MINSA 1996; and Petrera and Seinfeld 2007, ch.1.

[2] Since 2002, regions became legally responsible for the administration of health facilities in the country, reducing MINSA's administrative duties. Regions gained control over these facilities gradually throughout the last decade.

Arroyo 2000, 91–123; interviews with Arias Stella 2007, García 2007, and Tejada de Rivero 2007).

Expertise has also been very low in the health sector. Traditionally, politically connected doctors without a background in public health served as ministers of health for very short terms (Arroyo 2000; interviews with Johnson 2007 and Sánchez Moreno 2007). By my count, only five of twenty-five ministers from 1980 to 2013 had specialized training in public health. Ministerial turnover was very high, and the appointment of a new minister usually led to other personnel changes in the ministry (interviews with León García 2007, Vera 2007, and Pastor 2007). Apart from a period of stability during the nineties, most directors and advisers in the ministry occupied their positions for merely one ministerial term (Arroyo 2001, 50; interviews with García 2007 and Peña 2007). Significant changes in this policy area were very limited during the eighties due to worsening economic conditions that drastically reduced the sector's budget and strong resistance from health unions to even minor reforms (interviews with Arias Stella 2007, García 2007, and Pinillos 2007).

An exception to this immobility was the failed reform effort of Alan García's first health minister, David Tejada (1985–1987). A lifetime member of APRA, the incumbent party, Tejada was the most prestigious and influential health policy expert in the country. At the time of his appointment, Tejada was deputy director of the World Health Organization (WHO) and had helped organize the highly influential 1978 Alma Ata conference where WHO promoted primary health care (Cueto 2004). García presented Tejada's appointment as evidence of his commitment to the health reforms urged by WHO (interview with Arce 2007).

Once in his position, Tejada assumed that he had the necessary political support to implement ambitious changes, especially when his proposals received unanimous support in Congress (interview with Tejada de Rivero 2007). These proposals involved a drastic decentralization of material and human resources (Tejada de Rivero 1985). However, it seems that the government was not aware of the implications of taking primary health care seriously. Directing resources to rural areas was not popular among politicians or influential factions within the Ministry and led to confrontations with unionized doctors, nurses, health officials, and other health care experts (interviews with Arce 2007 and Tejada de Rivero 2007).

According to Tejada, another major impediment to his reform plan was MINSA's low level of expertise. The ministry lacked professional cadres that could implement primary health care–oriented reforms. In a desperate effort to compensate for this deficiency, Tejada himself started giving informal early morning classes to his directors and advisers (interviews with Arce 2007 and Tejada de Rivero 2007).

Overwhelmed by the protests, and lacking support from politicians who wanted to avoid clashes with unions and reform critics, Tejada left the ministry in 1987. He was reappointed in May 1989, but unions would not even give him the benefit of the doubt. More strikes amid the economic crisis forced his resignation only five months later. Tejada's failed effort to reform the health care system shows the severe challenges reformers faced at MINSA. Speaking playfully about his experience as minister, Tejada told me that although he knew many things about heath policy, he realized he would have needed an MA in political science to deal with the political and particular interests in the health sector.

Conditions in the health care system became even worse by the end of the decade (Vidal 1994a, 319–324). The economic crisis and political violence damaged the health care network, especially in rural areas (Ewig 2004, 221; interviews with Meloni 2007 and Luján 2007). Inflation reduced salaries to pittances, forcing MINSA professionals to work in private practice in order to make ends meet (Webb and Valencia 2006). Public hospitals implemented fees to fend off resource shortages (Arroyo 2000, 59–62). Corruption skyrocketed and the black market for medicine grew significantly (interviews with Espinoza 2007, Cubas 2007, and Meloni 2007). It was under these precarious conditions that health experts arrived at MINSA.

THE EMERGENCE AND STABILITY OF HEALTH EXPERTS (1990–2002)

In 1990, President Alberto Fujimori appointed Carlos Vidal as his first Minister of Health. A doctor and former vice president of the Pan American Health Organization (PAHO), Vidal recruited a cohort of young public health experts to fill MINSA's upper echelons. Many of these experts had MAs in public health from the prestigious *Universidad Cayetano Heredia* and even more had received technical training from PAHO or special programs supported by the University of Alabama, Johns Hopkins University, and prestigious Latin American institutions (Vidal 1994a, 316–317; Vidal 1994b, 345; interviews with Cubas 2007, Gozzer 2007, and Meloni 2007). These young experts contrasted with the older bureaucratic cadres in charge of the Ministry during APRA's last months in power.

Vidal's ambitious reform plan aimed to decentralize the health care system and increase preventive care (MINSA 1990). Technocrats appointed in the International Cooperation Office (*Oficina de Cooperación Externa*, or *OCE*) sought loans to increase the ministry's limited budget and initiate reforms (interview with Meloni 2007). These experts were aware of IFIs' and bilateral agencies' growing interest in health policy reform and whose support was a necessity with the ministerial budget under incredible stress (interview with Cubas 2007). Vidal also appointed two highly reputed experts to lead special programs for immunization and TB. Under severe constraints, the team designed a five-year plan (*Políticas y Plan de Salud 90–95*) that proposed significant changes to the Ministry, including merging it with the social security institute to build a single national health system (MINSA, 90–95, 1990a).

Nonetheless, this reform process came to an early halt when Vidal and Fujimori clashed about the proper way to handle a cholera epidemic in January 1991 (Reyna and Zapata 1991). An anecdote shows these disagreements. Vidal advised the population not to eat *ceviche*, a popular dish based on raw fish that was suspected to be contaminated with the bacteria due to sewage runoff from Lima. As a result, fishermen and some restaurant owners protested, alleging that there was no evidence that raw fish transmitted cholera. Some ministers showed their support for these protests by eating *ceviche* in public. Ultimately, President Fujimori himself ate *ceviche* on television to tell the population that it was perfectly safe, which undermined the credibility of his health minister (Vidal 1994a, 332–339). In February, Vidal resigned along with part of his team. However, other team members, such as personnel in the OCE and those experts leading the previously mentioned special programs, remained in their positions (interviews with Cubas 2007 and Meloni 2007).

Técnicos' expertise – and some fortunate events – allowed them to remain influential actors in MINSA from 1991 to 1993 by becoming ministerial advisers. After Vidal, Fujimori appointed Víctor Yamamoto, his sons' pediatrician, as minister. A doctor with no public health expertise, Yamamoto relied heavily on experts' advice to avoid mistakes (interviews with Espinoza 2007 and Meloni 2007). Leadership of MINSA changed again in November 1991 when Victor Paredes, a congressman from the incumbent party, was appointed minister. Paredes was known for his opposition to Carlos Boloña and the MEF's neoliberal policies. This rivalry helped experts maintain influence in the ministry, as they became Paredes' allies in opposition to the MEF's proposed health reforms.

In fact, the same month of Paredes's appointment, Boloña and MEF technocrats successfully pushed for a presidential decree enacting an ambitious neoliberal health policy reform project (Legislative Decree 718). With the goal of increasing efficiency and competition in the public health care system, the project took steps toward privatizing health care provision in Peru. In line with the Chilean health care reform model, the decree allowed workers to leave the social security institute and join private health care insurance groups called *Organizaciones de Servicios de Salud* (Weyland 2006, 185–188; Ewig 2010, 106–109). Experts in MINSA strongly disagreed with this reform due to what they believed were excessively high social costs (Ewig 2004; interview with Meloni 2007). Paredes used experts' arguments to convince Fujimori that the reform carried negative social and political consequences, leading to a delay in its implementation.

In August 1993, Congressman Jaime Freundt was appointed to head MINSA. A proponent of neoliberal health policy reform in Congress, Freundt was expected to support the MEF's plan. Nonetheless, the policy remained shelved (Weyland 2006, 187–188) and, surprisingly, Freundt retained some of Paredes' advisers. It was during Freundt's tenure that *técnicos* became powerful actors at MINSA. Two factors explain this decisive expansion of technical autonomy.

First, IFIs and bilateral aid agencies approved the loans negotiated in 1990 and 1991, pending some minor clarifications. These revisions required technical knowledge of similar reforms and the capacity to communicate with IFIs on technical matters. The minister gave more resources to the OCE in order to make these clarifications and assure the loans. According to the then-OCE director, because the reforms needed a more technical approach and less rigid administrative structure than the MINSA bureaucracy could provide, the decision to support experts in the OCE was mainly for efficiency's sake (interview with Meloni 2007),

These loans were finally approved, and three externally funded projects started that year (interview with Meloni 2007).[3] One of the projects, funded by the IDB and the Japanese cooperation agency (OECF), was used to strengthen the OCE (*Programa de Fortalecimiento de Servicios de Salud, PFSS*). The OCE administered this project and monitored two others that functioned as parallel bureaucracies within the ministry: the *Programa de Salud y Nutrición Básica* (*PSNB* and *Proyecto 2000*), which were financed mostly by the WB and USAID, respectively (Ewig 2010, 97–105; interviews with Meloni 2007 and Gozzer 2007).

A second factor that further contributed to experts' autonomy was the start of a new poverty alleviation campaign. With a stable economy and the Shining Path in retreat, the government outlined a plan to reduce poverty and compensate for the social costs of market reforms. The minister of economics presented the plan to creditors in the Paris Club, who approved it enthusiastically due to its embrace of recent IFI proposals for human development (Ewig 2010, 80–81). As a result, in 1994, MINSA started a targeted program to provide a basic health package to the poorest sectors of society (*Programa Salud Básica para Todos*, PSBT), fully funded by the national budget (Ewig 2010, 97–100). Designing and implementing PSBT also increased the need to hire experts in MINSA.

In what follows, I focus mostly on experts at the OCE, as this office became a think tank for health reform and lent support to other reform initiatives. However, there were other experts working in MINSA at the same time, some of them quite independently from one another. Although the externally funded programs were designed and initially supervised by the OCE, they eventually gained considerable independence from the office. Additionally, other initiatives like PSBT, the immunization and TB programs, and efforts to delegate the administration of medical posts to local communities, had special offices within the ministry that only minimally coordinated with the OCE (Ortiz de Zevallos et al. 1999; Ewig 1999: 508–509; Ewig 2004; interviews with Meloni 2007 and Vera 2007).

[3] The projects' budgets amounted to 220 million dollars, out of which 152 million were foreign aid funds and the rest governmental funds (Ugarte and Monje 1999).

This reform period was unique in Peru's history. Never before had MINSA had so many resources from foreign aid and so many health experts working in the ministry; as one official put it, the OCE was a "ministry within the ministry" composed of an "army" of health experts (interview with Luján 2007). Although, in contrast to the comprehensive overhaul of the health care system in Colombia, reforms were only "piecemeal" (Ewig 2004), they constituted unprecedented changes to Peruvian health care that were implemented by autonomous experts within the ministry. What evidence is there that experts were acting autonomously?

First, experts successfully implemented reforms to advance their preference for improving health care for the poor, especially in rural areas (similar reforms to the ones that politicians had been unwilling to support in the 1980s). To do so, experts deliberately shielded policy making, which allowed them to pursue their reforms without attracting political attention. *Técnicos* used this insulation to channel international and national resources toward providing selective primary health care to vulnerable groups of the population, especially in rural areas. Each project was directed toward these goals. The PSNB, for example, targeted infant and maternal care while also training health officials. (Ewig 1999; 2004; 2010; Petrera y Seinfeld 2007, 79–84; Ugarte y Monje 1999, 575–576). *Proyecto 2000* started pilot projects in five poor regions of the country and also provided training to health officials. Consequently, they reduced some of the urban and middle class bias of the Peruvian health care system (Ewig 2010, 56–58).

Additionally, experts privileged the creation of more primary care health centers. The cornerstone for a decentralized health system was the construction of primary care centers in shanty towns and rural areas. PFSS used a considerable portion of its funds to strengthen the health care network with new facilities, buy modern equipment, and improve its administrative capacities (interviews with Gozzer 2007 and Pedro Mendoza 2007). As seen in Table 7.1, while the number of hospitals remained constant during that decade, there was an explosion of new health care centers and primary care centers that clearly shows this change in priorities (Ugarte and Monje 1999, 575).

Additional evidence of autonomy is the fact that experts achieved considerable continuity and stability in a ministry characterized by high ministerial turnover, which was in turn accompanied by frequent personnel changes at lower levels. After Vidal, Fujimori appointed seven health policy ministers during the decade. Experts remained influential under all of them, even though, as the next section details, their relations with ministers were not always peaceful. However, their actions show that they were able to escape political control to a considerable degree.

Technocrats also were autonomous from IFIs, even though foreign resources were critical for the development of this technical bureaucracy. Experts had a close relationship with IFIs and followed some of these institutions' recommendations for health policy reform, such as promoting efficiency and

TABLE 7.1. *MINSA Health Infrastructure (1980–2002)*

	1980	1985	1990	1992	1996	2002
Hospitals	114	123	130	148	136	138
Health centers	451	592	826	826	1028	1179
Primary care centers	1402	1776	2662	2958	4762	5415
Other				2	7	111
Total	1967	2491	3618	3934	5933	6843

Source: MINSA 2007.

targeting vulnerable groups. At the same time, Peruvian experts had their own preferences that were not always consistent with IFI recommendations. *Técnicos*' preferences and policies also drew influence from the World Health Organization (WHO), the Pan American Health Organization (PAHO) and the United Nations Population Fund (UNFP), institutions that lack the kind of leverage attributed to IFIs. More importantly, there is no evidence of IFI coercion in the adoption of experts' reforms (Ewig 2010, 83).

Although these health policy experts did achieve significant autonomy, they were more limited in advancing their preferences and preventing political interference than were economic technocrats and their counterparts in the Colombian health sector since the 1990s. On the one hand, experts lacked the leverage to achieve the more comprehensive reform they desired. On the other hand, technocrats failed to resist political intervention in technical policies and eventually lost their relevance in the ministry.

Crucial evidence of OCE experts' limited autonomy is their failure to achieve comprehensive reform in line with their preferences. The document *Lineamientos de Política de Salud (1995–2000)* contained a plan to fully overhaul the health care system by balancing equity, efficiency, and quality (MINSA 1996; Arroyo 1999, 549; Arroyo 2001, 14). Several follow-up documents provided additional details for this proposal and developed a more precise diagnosis of the health care system (Mesa-Lago 1996; Espinoza 1998). The reform mirrored several aspects of the Colombian health policy reform, including a plan to provide subsidized health insurance for the poor. Equity was a principle guiding the reforms and, in comparison to MEF plans, the OCE's proposals relied less heavily on market mechanisms for the health sector. Although experts advocated for these reforms and gained the support of Minister Eduardo Yong Motta (1994–1996), the plan was never implemented due to its perceived political costs (Weyland 2006, 161, 191–192). The appointment of a more neoliberal minister in 1996 meant that MINSA changed priorities and shelved experts' comprehensive proposal.

Peruvian health policy experts also had difficulty limiting political interference in their ongoing policies. Two episodes clearly show these limitations. As just mentioned, in 1996 a new minister, Marino Costa Bauer, led a new effort to privatize social security health care in a similar vein as the failed reform of 1991 (OSS). The minister hired a law firm to review the 1991 OSS project and draft a new bill (Arroyo 1999, 14; Ewig 1999, 509–511). MEF experts supported the creation of private health care providers (*Empresas Prestadoras de Salud, EPS*), which the MINSA technocracy once again opposed (Ewig 1999, 510). Nonetheless, contrary to experts' preferences, the decree was finally approved. Although Congress eventually reversed this decree and the incumbent approved a considerably less comprehensive law, the fact is that experts were not able to block this reform. The proposal was watered down only because of its perceived political costs (Weyland 2006, 193).

A second political intervention occurred in July 1997 when President Fujimori announced the creation of a free insurance program for students (SEG). The president made the announcement during a message to the nation without consulting MINSA. As a result, the minister of health had to implement this insurance plan hastily without proper evaluation (Jaramillo and Parodi 2004; Arroyo 1999, 575; Ewig 2010, 112–114). Health policy experts opposed SEG, because they did not consider students a particularly vulnerable group. Limited resources, they argued, would be better spent on infant or maternal care. However, the decision was politically very attractive because it would benefit a broad section of the population with a comparatively low cost due to students' relative good health (Jaramillo and Parodi 2004, 16). In 1999, the ministry started a maternal and infant insurance program (*Seguro Materno-Infantil*) while maintaining SEG.

These political changes and clashes led to the gradual but steady decline of technocratic influence. In August of 1999, vice minister of Health Alejandro Aguinaga became minister. Aguinaga had been vice minister since 1994, during which time he had clashed with the OCE (anonymous interviews with two members of the OCE reform team). Due to his long tenure, Aguinaga understood health policy and how the ministry worked, so experts had less leverage over him.

Resources also became scarce, partly due to IFIs' diminishing interest in health policy reform and partly due to the change of government. As mentioned in the previous chapter, IFI emphasis on social policy decreased by the end of the decade. The IDB, for example, did not renew its loan for PFSS. OCE started to seek funds for a new project called PARSALUD (*Programa de Apoyo a la Reforma de Salud*), which would have allowed MINSA to adopt a more comprehensive plan of public insurance for vulnerable sectors and maintain high levels of expertise in the ministry. The loan consisted of $270 million (U.S. million) from foreign donors, with additional money from the national budget. The abrupt end of Fujimori's government in 2000 caused international lenders to reconsider these loans (MINSA 2001, 23) and, during the transition government, the MEF opted not to pursue them for budgetary reasons. A smaller loan of $88 million was finally approved that year for PARSALUD (interviews with Meloni 2007 and Pretell 2007).

Additionally, the new administration restructured the OCE (MINSA 2001, 28). Some of the new authorities criticized the office, and the foreign aid projects in general, for developing parallel organizations in MINSA ("Islands of Efficiency") without using their resources to reinforce MINSA's core bureaucratic capacities (MINSA 2001, 21–22; interviews with Pretell 2007 and Vera 2007). This shortage of resources, along with internal disputes among experts concerning their participation in Fujimori's reelection campaign, led to the departure of several *técnicos*.

The end of this health policy technocracy came under President Toledo's government in 2002 when Luis Solari (July 2001–January 2002), a doctor and leading member of President Toledo's political party, *Perú Posible*, was appointed minister of health. A devout Catholic, Solari opposed several policies promoted by the ministry, especially family planning programs. According to Solari, many experts who had worked in the ministry under Fujimori were still in their positions when he reached the ministry. Solari accused them of implementing abusive health policies, such as a controversial plan in which patients were sterilized without proper consent. Similarly, in his opinion, they were involved in corruption within the ministry (interview with Solari 2007). Solari's successor, Fernando Carbone (January 2002–June 2003), shared these convictions and policy preferences. Diverse experts recall this "Catholic" period as one when several experts left the ministry, affecting the continuity of ongoing programs (interviews with Espinoza 2007, Luján 2007 and Meloni 2007). Priorities, personnel, and policies were easily changed at MINSA.

Since then, no similar technocracy has emerged at MINSA, although there have been several more technically oriented ministers. In October 2008, Alan García appointed a respected health policy expert, Óscar Ugarte, as minister of health. His longer tenure (2008–2011) allowed for some policy continuity and technical reinforcement of certain areas in the ministry. Ollanta Humala's second minister, Midori de Habich (2013–) is also a widely known expert and has plans for ambitious reforms. Nonetheless, it is too early to tell if she and her reform team will gain continuity and achieve significant autonomy.

In conclusion, experts made meaningful changes to the Peruvian health care system and achieved autonomy for almost a decade. Some of their policies and reform agendas still are considered controversial, but their leverage in the sector is widely recognized. Nonetheless, this episode also shows the limits of health policy experts' autonomy and the difficulty of entrenching this autonomy. What explains this period of moderate autonomy that ended in the experts' demise?

EXPLAINING HEALTH EXPERTS' LIMITED AUTONOMY AND EVENTUAL DEMISE

This technocratic episode at MINSA highlights the effect of changes over time in some of the factors in my theory. As in the previous episodes, experts gained autonomy when politicians appointed them to implement policies that neither they nor the ministerial bureaucracy could enact. Intermediate scores in two of the factors favored this outcome: the increasing complexity of health policy economics and the involvement of IFIs in health policy that evened the balance of stakeholders in the health sector. The two other factors, technical consensus and the political cost of poor policy performance, remained low. Thus when IFI interest declined and politicians clashed with experts, technical autonomy declined. Peruvian reforms did not incorporate middle and upper classes into the public health system nor did they include private stakeholders interested in public-good provision. Thus, these reforms did not have the same effects as those in Colombia that made politicians concerned about health policy and reliant on experts in the health sector. In this section, I first discuss how experts exploited their knowledge to enhance technical autonomy; then I focus on the four factors of my theory, looking at which favored technical autonomy and which limited it.

For almost a decade, until their demise, experts carefully maneuvered to avoid political clashes and advance their reforms. As discussed previously, politicians appointed experts and provided them with significant power to address a series of external and internal demands. This delegation was significant, allowing experts to work on their reform projects independent from political oversight, but members of OCE also crafted this autonomy by acting strategically. An example of this strategic action is experts' secrecy. In contrast to Colombian experts who publicly discussed the content of reforms, their Peruvian counterparts opted to form policy proposals among themselves, with little external involvement. This secretive style explains why Peruvian health policy making was called "politics of stealth" (Ewig 1999; 2004).

The technical team was aware that achieving legal changes in Congress required negotiation and the active involvement of politicians, which might arouse opposition. To avoid such problems, most of these reforms were made through presidential or ministerial decrees. For example, PSBT, CLAS and other programs were all adopted through such infra-legal rule changes, without congressional involvement (Ewig 1999, 490–493; Arroyo 1999; interview with Meloni 2007). Laws to institutionalize some of these reforms were adopted much later, in 1997, with the approval of the *Ley de Modernización de la Seguridad Social en Salud* (Ley 26790) and the *Ley General de Salud* (Ley 26842) (Arroyo 1999, 542–547; Ewig 1999; see also Ortiz de Zevallos et al. 1999, 22).

The success of this "silent" strategy shows how much leverage experts can gain when politicians lack information about ongoing policies. As discussed in previous chapters, incumbent politicians were not fully aware of the reforms and left decisions to the experts. Moreover, as discussed by diverse authors, political opponents also remained unaware because they paid little attention to health sector reforms and were not informed about changes in the ministry (Arroyo 1999, 549–551; Ewig 2010, 98; Ortiz de Zevallos et al 1999, 27).

Commentators have suggested that this strategy made the policy changes more vulnerable to future reversal (Arroyo 1999; Ewig 1999; 2004, 242–243; Francke 2001; Ugarte and Monje 1999, 606–611; interviews with Francke 2007 and Pretell 2007). By not consulting with other actors, some argue that experts failed to build a broad constituency in favor of reform that could have prevented reversals. Others have highlighted the potential negative effects of not debating publicly the content of these changes. Nevertheless, writing policy behind closed doors was successful in the short term and clearly shows experts' independent decision making.

Further evidence of experts' strategic action was their use of IFIs to protect their reform efforts. Experts sought IFI support to prevent political intervention in the reforms. For example, experts asked IFIs to pressure the government to modify Fujimori's school insurance program (SEG) so it would complement the maternal and infant health insurance program that was later adopted (Weyland 2006, 161–162). Similarly, they also supported IFIs' and bilateral agencies' decision to insulate reform projects from the ministerial bureaucracy. Experts were aware that this insulation reduced the chances of political intervention, enhanced efficiency, and guaranteed autonomous room for expertise.

However, as discussed in previous chapters, experts' strategies are only part of the story. Two factors identified by my theory – complexity and a more balanced constellation of stakeholders – explain experts' autonomy and relative success in advancing a reform agenda. The low scores in the remaining factors explain the limits to this autonomy and the experts' final demise. I first discuss the two positive factors, and then the remaining two that scored low.

Positive Factors

With the growing interest of IFIs, such as the World Bank and the IDB, in health policy during the nineties, a more plural and balanced constellation of stakeholders temporarily emerged. This episode exemplifies how a change in the constellation of stakeholders can provide an opening for expertise to emerge. This more balanced constellation of stakeholders favored experts in many ways. First, it contributed to the appointment of experts. The incumbent appointed experts to poverty-alleviation programs in 1993 in order to show the international financial community that the government was adopting sound, responsible policies. Further, as mentioned, experts had the necessary knowledge to respond to IFI demands and achieve the loan projects approved in 1994.

Second, it made experts necessary for politicians unwilling to alter the reform projects already under way. Ministers were aware that replacing technocrats could affect IFI sponsored reform projects, providing experts with an indirect coat of protection. In that way, IFIs gave experts more political space and resources to advance their preferences and achieve autonomy. Finally, as discussed previously, experts strategically used IFIs to gain leverage against the incumbent. For all these reasons, IFIs' involvement enhanced technical autonomy by making a more balanced constellation of stakeholders surrounding health policy.

However, even including IFIs, this constellation of stakeholders still remained unbalanced. To begin with, IFIs by themselves lack the power to check the incumbent's power and limit intrusion in health policy, particularly in noncrisis conditions. IFIs' involvement is always limited. These institutions may oppose certain decisions and exercise leverage by withholding funding, but they cannot block politicians determined to intervene in a certain policy area. For example, although IFIs opposed the SEG program, they could not stop Fujimori from going ahead with the plan. No expert interviewed mentioned IFI support as important to maintaining technical expertise when under political pressure. Two *técnicos* interviewed, who prefer to remain anonymous, even complained bitterly about the lack of support from these agencies when conflicts with politicians arose.

Furthermore, other than IFIs, no powerful actors were interested in health policy. There were only a handful of NGOs and some think tanks that paid attention to the health sector while private actors remained largely apathetic. Additionally, in open contrast to the situation in Colombia, the reforms did not foster any new stakeholders with strong interests on the content and quality of health policy. Despite its expanded coverage and new programs, the reforms largely maintained the structure of the system that had existed before the 2000s. Thus when IFI involvement in health policy reform declined after 2000, the constellation of stakeholders became even more unbalanced.

Increasing policy complexity due to health policy economics contributed to this period of expert autonomy. Now policies had to meet new criteria, such as cost effectiveness, which made decision making much more complex. This complexity contributed to experts' autonomy, as the two nontechnical ministers who followed Vidal after 1991 were forced to rely on advice from technocrats. Policy complexity also explains technocrats' considerable autonomy after 1993. It was the need to have experts in the OCE who could effectively deal with international agencies in order to obtain external loans, and the decision to implement complex, targeted, poverty-alleviation policies that gave experts in MINSA considerable leverage, even if some ministers, such as Freundt or Costa Bauer, did not fully share their reform agenda.

Handling these complex reform projects gave experts informal protection from political interference. Advancing reforms demanded high technical capacities to communicate with IFIs, which many politicians lack. According to the experts interviewed, each new minister wanted to know about the OCE projects. Once they learned what OCE was working on, ministers realized that interfering in the office would potentially affect these complex reforms (interviews with Meloni 2007 and Gozzer 2007). As a result, ministers preferred to leave the status quo untouched. Three different members of the OCE told the same anecdote. Given the usual short tenures in the health sector, these experts assumed that when they were first hired, their appointment would be brief. After some months, they started to realize that they had achieved considerable stability, even more so than the health ministers who "came and went" while experts stayed in their positions (interviews with Gozzer 2007, Meloni 2007, and Mendoza 2007).

This stability is explained by the fact that experts were in charge of complex, ongoing reforms, which made dismissing them politically risky. As discussed later, this hesitation to intervene with internationally backed projects is also evidence of the importance of stakeholders, such as IFIs, for technical autonomy. In sum, the higher level of technical complexity in this policy area, partly caused by health policy economics, increased the level of expertise required to design and implement policies.

Nonetheless, even if more sophisticated than before, health policy has remained less complex than economic policy. The consequences of these reforms are more understandable for those working, or having interests, in the health sector. Several of these decisions involve the allocation of distributable goods across the territory, something that politicians are well aware of. This lower complexity allowed other actors, such as doctors or politicians, to interfere in policy decisions and reduce the amount of autonomy experts had. In particular, Peruvian reform efforts were sometimes thwarted by political interests. The clearest political intervention during the decade was Fujimori's decision to adopt the SEG against experts' advice. Experts thus remained more vulnerable in this important respect.

Negative Factors

The remaining two factors score low, thus helping us understand this episode's lower technical autonomy and the difficulties of entrenchment. First and more important, this episode confirms that delivering poor health policy outcomes has low political costs. Those affected by ineffective health policies in Peru are generally poor and command very limited political influence. Additionally, the population has grown accustomed to poor health policy and does not link poor public health care to the incumbent.

As a result, health policy is not a major political issue. The topic of health care rarely comes up during presidential campaigns, and when it does, it usually involves vague promises of reform with little technical substance. Evidence of this lack of public interest in health policy is visible in opinion polls. In recent years (2008, 2009, 2010, and 2011), when asked what were the three most serious problems in Peru, poor health policy (*salud pública inadecuada*) ranked ninth.[4] In a similar question, in 2013, health ranked

[4] Ipsos-Apoyo, Opinión Data, Year 10, N.123. April 19th 2010. http://www.ipsos-Apoyo.com.pe/sites/default/files/opinion_data/Opinion%20Data%20Abril%202010.pdf; 28th February 2011, http://www.ipsos.pe/sites/default/files/opinion_data/Opinion%20Data%20Febrero%202011.pdf.

twelfth.[5] Asked about what issues candidates should address in debates during the presidential campaign, Peruvians ranked health seventh in 2006.[6] Even though other polls with a narrower list of issues do reveal greater public concern with health care, it rarely becomes politicized (IOP 2013, 5). As a result, there are few political consequences if politicians intrude in ongoing reforms and if performance in this sector remains weak.

As previously discussed, this lack of political interest sometimes favored experts, as they could advance policies with little interference from politicians who were unaware of the ongoing reforms. Yet it also reduced the urgency of conducting reforms, which limited experts' leverage. And when politicians did interfere, they could do so without political cost. Removing technical ministers and appointing prestigious doctors with little public health knowledge did not receive much public attention. Sometimes, politicians can even achieve short-term popular support by intervening. President Fujimori's decision to adopt SEG in 1997 provides a good example: the policy was electorally attractive for his reelection effort in 2000. Ministers' interference with experts' policies or their lack of interest in technocratic reform plans clearly shows the relevance of this factor.

In contrast to the Colombian case, Peruvian reforms did not change the system drastically in order to incorporate middle and upper classes. The formal-sector workers maintained a separate social security system that provided health care, and the MEF-sponsored reforms created several private exit options. This continuing detachment of the middle and upper classes from the health system reduced the political saliency of this issue in Peru.

The first factor related to the technical aspects of policy – expert consensus – also remained low in the Peru case. This was paradoxical. By the time of the reforms, the international health community was already reaching a consensus around a broad version of selective primary health care (Cueto 2004, 1868–1869). Although selective primary health care was not yet an entirely unified paradigm, IFIs and a variety of international health organizations, such as the PAHO, WHO, and UNICEF, gradually took interest in selective primary health care, partly because it proposed realistic, measurable, and easy to monitor strategies (Yglesias 1999, 531).

Health experts in the OCE considered themselves part of this growing international consensus. In the words of one expert, "We are in a process of internationalization of the politics of health. Like in economics, you see that there are a series of economic policies that have an international level of consensus. Health [policy] is in a process of that kind" (anonymous health expert, interviewed by Ewig 2010, 81). Experts' designed their 1995 reform plan adhering to this consensus. Technocrats sought to promote both equity through subsidized insurance regimes and efficiency by creating competition

[5] Ipsos-Apoyo, Opinión Data. April 2013.

[6] Ipsos-Apoyo, Opinión Data. May 2006.

in the health and social security systems. According to a local observer, this reform plan represented a centrist option (Arroyo 2001, 14).

Nonetheless, within Peru, this nascent consensus did not exist in the early nineties. Some Peruvian health policy experts saw the international position as too neoliberal. These *técnicos*, who favored a single national health care system (Arroyo 2000, 73), were active during the nineties in the Peruvian Academy of Health, medical associations, and the Institute of Social Security (interview with Sánchez Moreno 2007). Another group of experts continued to promote the original version of primary health care or a rights-based approach to health policy (interview with Tejada de Rivero 2007; Arroyo 2000, 75–76, 161; Yglesias 1999, 529; Vidal 1994a, 314). These specialists accused experts in MINSA of uncritical foreign health policy reform recipes.[7]

Yet MINSA experts were vulnerable to proposals from the right. MEF technocrats, in particular, favored an extreme neoliberal version of selective primary health care inspired by the Chilean health policy reform. As evidenced by the MEF's support for the neoliberal proposals contained in the OSS decree of 1991 and the EPS reform of 1996, these policy divisions remained acute throughout the decade (Arroyo 1999, 552–553; 2000, 162).

This lack of consensus over health policy weakened experts' position in favor of a centrist reform. As discussed in Chapter 3, this lack of consensus opens the door for politicians to choose among these different available options. The neoliberal experts behind Boloña's OSS and Costa Bauer's EPS reforms show how divergent technical opinions could be exploited by politicians to advance their preferred policies. Similarly, when Fujimori demanded the implementation of SEG, the minister appointed experts willing to do so (Ewig 2010: 112–115).

As these incidents show, experts with different perspectives were sometimes working side by side within MINSA but pushing reforms in different directions. Internal disputes wasted energy that could have been directed toward a common reform plan. MINSA experts certainly did not help build a consensus on health policy in Peru; they opted not to consult other actors in the health sector or publicly debate reform proposals, which may have reduced the possibility of achieving a broader consensus (Ewig 1999:491–492). Clearly, this lack of consensus did not allow reformist health policy experts to present a unified front to politicians, thereby weakening their position.

The medium scores – at least temporarily – on two of the factors in my theory, and the low scores on the others, indicate that experts' autonomy was weak and more vulnerable to political intrusion. Experts enjoyed considerable

[7] Ministers themselves tend to have quite strong opinions about the perspectives preferred by their predecessors. For example, Minister Vidal criticized strongly Minister Tejada and his "health for all" approach for its strong emphasis on preventive measures. According to Vidal, such emphasis leaves aside curative measures and disregards the importance of hospitals for public health (Vidal 1994a, 314–315).

leverage in MINSA, but never achieved enough power to push through their preferred policies or gain a stable, entrenched position. When IFIs' interest in health policy reform declined and experts clashed with politicians, technical autonomy in MINSA largely evaporated.

Before concluding, it is worth mentioning the MINSA epidemiology office, which has been an exceptional pocket of autonomy over the last thirty years, has maintained high levels of technical expertise and continuity. The personnel in this area have more technical skills than elsewhere in the health sector bureaucracy, and its directors enjoy longer tenures (Arroyo 2001, 50; interviews with Espinoza 2007 and Ramírez 2007). This office is part of an international community of epidemic research, receiving funds from USAID and Johns Hopkins University to collect and analyze information about epidemics in Peru.

This higher technical level was successfully crafted by epidemiology officials. To insulate and professionalize the office during the eighties, epidemiology officials created a special course on epidemic readiness and response (*Programa de Especialización en Epidemiología de Campo,* or PREC) that all personnel were required to take (interviews with Espinoza 2007 and Ramírez 2007). The program was fully funded by external aid and administered by a private university, thus limiting political influence. PREC shows how epidemiologists successfully reformed their own rules around appointment to guarantee that only qualified individuals worked in this area. Why were epidemiologists capable of doing this?

My theory shows how more favorable conditions in this area led epidemiologists to have greater autonomy. First, policy making in the area is quite complex. Officials require specialized education in controlling epidemics and must be capable of managing a complex statistical system that tracks epidemics, which limits the possibility that unqualified personnel are appointed. Second, epidemiology is an area in which expertise remains largely uncontested and epidemiologists form a close, cohesive, and global community. Third, the political costs of not following specialized advice in epidemics control could be enormous. As mentioned by a health specialist, no politician wants to deal with a pandemic (interview with Espinoza 2007). These costs provided epidemiology with a sort of "natural" protection from political intrusion. Finally, stakeholders such as USAID and Johns Hopkins University support *epidemiólogos* with resources and are key allies to further their expertise. As a result, the office remains quite professional and autonomous. In sum, where several of my factors happen to be present, technocratic autonomy does emerge in the health sector.

CONCLUSION

Along with the previous chapter, the Peruvian health case shows that even in less likely policy areas, experts can muster considerable autonomy and advance their preferred policies. The need to reform health care led to the appointment of experts in MINSA. From the OCE and other areas of MINSA, experts employing various strategies were able to conduct piecemeal health policy reforms and insulate these changes from political intervention. The higher scores on two of the factors (policy complexity and constellation of stakeholders) contributed to the temporary autonomy of health experts in MINSA. Nonetheless, the lower scores in the remaining factors explain why this technocracy was not strong enough to resist intrusion and, as a result, eventually failed to entrench itself. The conditions were not favorable for technical autonomy, especially once IFIs lost some of their interest on health policy reform at the end of the decade.

A crucial difference with the more successful Colombian health technocracy lies in the lack of powerful stakeholders invested in keeping experts in charge of health policy after the reform. In Peru, health provision still remains linked to hospitals and other public facilities. There are few private actors involved in public health, such as health providers or insurers as in Colombia, that create cross pressures on the incumbent to appoint *técnicos* and then informally protect them. Without such stakeholders, Peruvian experts remained vulnerable and had fewer opportunities to entrench their autonomy.

Additionally important, in Colombia the health system now includes middle- and upper class individuals, thus increasing the political cost of poor policy performance. Colombian politicians face strong public criticism if they do not respond to crises in the medical system. Yet in Peru, the health care system mainly concerns poor sectors of society and is largely irrelevant in public debate except around the time of elections.

Interestingly, although these conditions in Peru overall remain unfavorable, the factors related to technical policy have now become more conducive to expert autonomy than in the past. In the last decade, the epistemic community of health policy experts in Peru has grown and health policy has become more complex. Since 2000, this community has helped the World Bank and local think tanks produce a considerable amount of work on public health (Alvarado Pérez and Lenz 2006, 145–147). Among these experts exists a growing consensus on what constitutes adequate health policy. Even if some experts promote a rights-based approach to health care and others focus more on selective health policies, most agree that any health policy reforms need to address resource limitations and prioritize reforms toward those most in need (Arroyo 1999, 561; 2000, 86; Chon et al. 2011; Francke 2001, 21–23; interview with Portocarrero 2010). However, until now these changes in the technical aspects of policy in civil society have not led to a significant increase in experts' autonomy in MINSA.

8

Conclusion: Technocratic Autonomy, Its Limits, and Democracy in Latin America

This book presents an in-depth comparative analysis of technocrats in Latin American democracies. Through the study of six technocratic episodes in Colombia and Peru, I propose that technocrats in the region are best conceptualized as autonomous, capable of pursuing preferences that are distinct from those of other important sociopolitical actors, such as their political superiors, IFIs, and businesses. This finding provides strong evidence against arguments that derive experts' role from these powerful actors. Instead, technocrats are independent actors who, due to their position in the state, are frequently first movers in policy reform thus successful in implementing their preferred changes. Experts' actions and choices carry important economic, social, and political consequences in Latin America.

The theory of technocratic autonomy developed in Chapter 3 explains both the source of experts' autonomy and the variation in this autonomy across policy areas and time periods. Technocrats' main source of power is their expertise. After their appointment, generally motivated by politicians' need to solve an acute problem, experts gradually gain control over policy making. By increasing the sophistication of public policy, experts immediately raise the technical bar required to participate in policy debates. Yet technocrats are more than passive actors: they use their expertise strategically, aiming to protect and enhance their autonomy.

The degree of experts' autonomy, however, varies across policy areas. Some experts are more influential and more likely to entrench their autonomy than are others. My theory explains this variation with four factors. The first two politics of policy factors determine where technocratic autonomy is more likely to emerge and become entrenched. The first of these factors is the political cost of poor policy performance; the second is the constellation of powerful stakeholders surrounding a policy area. When poor policy performance has severe negative consequences for politicians, and there is a balanced collection of stakeholders in which each prefers empowering experts over allowing another actor to control policy making, technical autonomy becomes more likely. Together, these two factors explain a considerable degree of the variation in whether experts enjoy meaningful autonomy and are protected from political intrusion.

By granting technocrats more space in which to develop their autonomy, these conditions also largely determine if experts will remain influential over time. Technical entrenchment occurs as citizens and powerful actors become invested in technical stability and thus support experts' continuity. Furthermore, the policies technocrats sponsor can raise the political cost of interfering with expert continuity and increase the influence of actors who prefer experts to remain empowered.

Although less consequential, two other factors related to the technical aspects of policy also explain some variation in technocratic autonomy. When positive, these factors increase both autonomy and the chances of technical continuity. First, greater technical complexity in a policy area makes expertise more necessary and helps experts to exclude nontechnical actors from policy making. Second, strong technocratic consensus in a policy area gives experts' prescriptions legitimacy and limits the possibility that sitting experts are replaced by others with preferences closer to those of the incumbent. These conditions are most relevant when the politics of policy factors already favor technical autonomy. By themselves, technical consensus and complexity will not get experts appointed in the state. And even if experts are appointed, these factors will not provide technocrats the leverage and informal protections enjoyed by those in policy areas where the first two factors are high. However, when the costs of policy performance and the constellation of stakeholders favor experts, these technical aspects of policy making significantly affect the level of technical autonomy and whether this independence entrenches.

The analysis of economic and health policy experts in Colombia and Peru provides empirical support for this theory. In all six episodes, technocrats gained considerable influence over policy making in areas of government under their control; even health policy experts exercised considerable influence despite existing theories that suggest this outcome is unlikely. Consistent with my theory, however, I show that economic experts enjoyed significantly more autonomy than did health policy experts. Economic technocrats were more likely to gain autonomy and entrench it over time, while health policy experts remained more vulnerable and entrenchment less likely.

In two of the episodes, Colombia from 1958 to 2013 and Peru from 1990 to 2013, economic technocrats exercised considerable autonomy and achieved long-lasting stability because of their strategic actions and the advantageous conditions they enjoyed. In the three health policy episodes, experts mustered enough autonomy to conduct reforms that shaped policy within their issue area. Nonetheless, as expected by my theory, in the two health policy episodes, Colombia from 1966 to 1970 and Peru from 1990 to 2002, in which the four factors scored lower, experts eventually lost their positions after clashing with political interests.

Variation in these factors also explains how autonomy within a policy area changes over time. In the first Peruvian economic episode (1980–1983), politicians did not realize the potential political costs of an economic crisis, so they continually interfered with technical decisions to cut public deficits and subordinated policy making to electoral interests. Additionally, the absence of a balanced constellation of stakeholders limited experts' autonomy. These factors largely explain experts' demise, as there were few barriers to political intervention in the policy area. Furthermore, the lack of technical consensus weakened technocrats, because politicians could pick other experts who would question the conclusions of "Team Dynamo" on substantive grounds. As a result, politicians and business interests were able to successfully prevent MEF technocrats from implementing their preferred reforms.

In contrast, higher political costs of poor health policy and a more plural constellation of stakeholders facilitated the emergence of a more autonomous public health technocracy in Colombia in the early 1990s that has persisted to the present day. The increasing complexity of health policy due to the introduction of health policy economics contributed to this outcome by widening the information asymmetry between experts and other actors. Although neither as strong nor stable as economic technocrats, Colombian health policy experts have remained relevant actors in health policy and, in the medium term, Colombian presidents seem dependent on them to manage a politically sensitive policy area.

Similar trends to those observed in my economic episodes are found in diverse Latin American countries, such as Argentina, Bolivia, Brazil, Chile, El Salvador, and Mexico, where economic policy is, or has been, dominated by technocrats (Koivumaeki 2010; Montecinos 2005; P. Silva 2008; Babb 2001; Schneider 1998; Teichman 1997). After their appointment to economic institutions in the 1970s and 1980s, economic experts achieved meaningful autonomy and significant stability. Although some technocrats eventually lost this power, others have maintained their relevance and capacity to overcome political opposition to their proposals, even where political turnover has produced new incumbents with different policy preferences than the politicians who initially appointed the experts.

In México, experts gained autonomy in the Ministry of Finance when the government of Carlos Salinas (1988–1994) sought to implement economic reforms (Teichman 1997; Centeno 1993). After Mexico's democratic transition in 2000, these technocrats remained strong. A high degree of technical control over economic policy also persisted in Chile after its 1989 democratic transition and a string of presidents from the center-left *Concertación*. Christian Democrats hired the so-called CIEPLAN monks to take over economic policy from Pinochet's Chicago Boys (Patricio Silva 2008). The socialist governments of Ricardo Lagos and Michele Bachelet, and right-wing President Sebastián Piñera, have all continued to select technically qualified appointees for economic institutions. Similarly, the leftist *Partido dos Trabalhadores* in Brazil was very careful to signal a commitment to economic policy continuity with their first ministerial appointments after the highly technocratic government of Fernando Henrique Cardoso.

At the same time, with exception of Brazil, Costa Rica, and a few other countries, health policy experts in Latin America are generally much less influential than are their economic counterparts and their stability more precarious. During the nineties, IFI-sponsored reforms expanded technical capacities in health policy across the region, but most ministries remain poorly financed and technically weak (Kaufman and Nelson 2004; Nelson 2004).

While the exceptions to this pattern deserve a more detailed analysis than I can provide here, the Costa Rican case provides some insights about the relevance of the politics of policy factors. This health bureaucracy has attracted academic attention for its quality and professionalism (McGuire 2010, 65–93; Weyland 2006, ch. 4). Although what we find in this country is more a professional health bureaucracy than a technocratic "island of efficiency" within a larger ministry, my explanatory factors are still relevant for this case. In Costa Rica, since the early seventies all social classes belong to a single health care system, which raises the costs of failing to deliver adequate health policy outcomes (McGuire 2010, 74–76; McGuire 2001, 1681). Additionally, strong parties competing for lower- and middle-class voters, and workers' unions (although weak) acting as stakeholders that wanted to strengthen social services, made health policy an important national political issue (McGuire 2010, 83–84; McGuire 2001, 1677). As a result, health policy experts have significant control over policy to the point that they can successfully oppose IFI prescriptions (Weyland 2006, ch. 4; McGuire 2010, 80–81).

More broadly, my theory's predictions can be applied to *técnicos* in other policy areas of Latin American states. Ministries of commerce and central banks are both policy making institutions with similar conditions to those in economic areas, especially in regard to my politics of policy factors. Thus, they are more likely cases for technocratic autonomy. In these policy areas, we observe considerably greater technical stability across the region.[1] In contrast, conditions in social policy sectors, such as education and labor ministries, resemble those found in health ministries; they are thus less likely areas for technical autonomy. Reform teams in these policy areas can sometimes enact significant changes (Grindle 2004; Nelson 2004), but fully implementing them and achieving continuity for experts are more difficult.

These results shed light on the understudied variation in capacity of different agencies within the state throughout Latin America and the developing world.[2] While some sectors inside Latin American states are highly professional, others

[1] The autonomy of central banks in the region are discussed in Taylor 2009 and Boylan 1998.

[2] For exceptions see: E. Huber 1995: 169–170; Hildebrandt and Grindle 1997; Graham et al. 1999; Bersch, Praça, and Taylor 2012.

remain poorly financed, badly equipped, and largely unable to achieve their institutional goals. Although capacity does vary between state institutions in developed countries as well, it tends to be more uniform, because agencies have a larger base of resources and capable personnel that allow them to perform their duties. The factors I identify open an avenue of research examining which agencies are more likely to gain autonomy and efficiently exercise their authority.

In sum, this book confirms the importance of technocrats as political actors in Latin America and their decisive role in policy making. I explain the source of technical power and its variation across both time and policy areas. Some questions remain open, however. Under what conditions would powerful experts that have achieved considerable continuity lose their power? What are the implications of these findings for theoretical debates about policy making and politics in the region? What are the practical lessons of these results for policy makers and reformers? Finally, do autonomous experts affect the quality of democracy in the region? In this concluding section, I focus on these and other issues.

THE FALL OF EXPERTS

My descriptions of economic technocrats as autonomous and entrenched may lead readers unfamiliar with Latin America to think that once they have established themselves in a policy area, experts are resilient to almost any political challenge. Recent episodes in the region show that this conclusion is exaggerated. Other economic experts, widely seen as powerful and stable in their countries, fell and policy areas became less technical. New officials appointed in these areas were subordinated to politicians' demands and lost considerable control over policy making. How and why did these entrenched experts lose their saliency?

As discussed in Chapter 3, I propose that the politics of policy factors from my theory are crucial for limiting political intrusion in technical policy areas and thus explaining technical downfall. Abrupt changes in these two factors open the door for technical demise even when technical autonomy is entrenched. Thus we expect events that reduce the political costs of firing experts, and political processes that drastically unbalance the constellation of stakeholders in a policy area, to lead to technocrats' downfall. Although these factors can operate independently, together they make the fall of *técnicos* more likely. Additionally, the lack of technical consensus can contribute to this outcome by weakening the legitimacy of state experts' technical prescriptions and allowing politicians to select other technically qualified appointees with competing perspectives.

Cases of economic technocratic downfall in the Argentine and Bolivian economic ministries show the relevance of these factors. Although not an economic case, the demise of technocrats in Venezuela's state oil company (*Petroléos de*

Venezuela, PDVSA) also demonstrates the saliency of these two factors. In these three episodes, we see an overall loss of technocratic autonomy, with considerable decreases in the technical level of the policy area and significantly more dependence on incumbents.

Throughout the nineties, technocrats in the Argentine Ministry of Economics were widely considered powerful and secure in their positions. Although the Ministry of Finance had long been more professional than were other areas of the Argentine state, it became highly technocratic and increasingly powerful during the neoliberal reforms under Peronist President Carlos Menem (Teichman 1997). Domingo Cavallo, Menem's second Minister of Economics, represented the apex of this technical power. Appointed to resolve the country's economic crisis, from 1991 to 1996 Cavallo oversaw a profound market reform process that drastically changed the Argentinean state (Teichman 1997; Weyland 2002a, 148; Corrales 2011). Among Cavallo's adopted measures was a bold and original convertibility plan that tied the Argentinean Peso to the U.S. dollar. IFIs and the business community considered Cavallo's continuity to be a clear commitment to prudent macroeconomic policy and fiscal responsibility that limited Menem's capacity to control the ministry (Teichman 1997, 48). After Cavallo, the ministry retained its influential position in Argentinean politics, and diverse authors have highlighted the immense power of economic technocrats during those years (Corrales 2011; Teichman 1997; Weyland 2002a). President Fernando de la Rúa (1999–2001), a politician from the opposition Radical Party, bowed to economic continuity, promising to maintain the market model and convertibility. Convertibility was perceived by the population and powerful economic actors as a guarantee of stability (Panizza 2013, 30).

This situation began to change with the 1999 financial crisis. Economic technocrats in the ministry were associated with the neoliberal reforms of the early nineties, which a growing percentage of Argentines believed contributed to the crisis. After failing to respond to the crisis, De la Rúa attempted to address the growing economic emergency by reappointing, in March 2001, Cavallo as his third Minister of Economics (Weyland 2002a, 202). However, this move proved unpopular with many Argentines who blamed Cavallo's reforms for their current problems (Weyland 2002a, 202–204).

Some of Cavallo's measures, such as his controversial decision to freeze bank accounts (*el corralito*), further polarized Argentines and, more interestingly, attracted criticism from stakeholders such as the international financial community (Weyland 2002a, 204). When the ongoing crisis led President De la Rúa to resign in December 2001, Cavallo also left the Ministry. After several caretaker presidents appointed by Congress resigned in the following month, Eduardo Duhalde was able to achieve some political and economic stability upon taking office in January 2002 and finish out De la Rua's term. One of Duhalde's first measures was to end convertibility.

Duhalde appointed Roberto Lavagna as Minister of Economics in May 2002. A heterodox economist who was part of former President Raúl Alfonsin's

(1983–1989) economic team, Lavagna changed some orthodox policies while maintaining quite strict macroeconomic management (Panizza 2013, 33). A change of technical perspective did not lead to a reduction of technocratic power. Newly elected President Néstor Kirchner (2003–2007), an old-style populist of the Peronist party, retained Lavagna in his position.

Nonetheless, when the crisis receded and Kirchner achieved a clear victory in the November 2005 legislative elections, the president asked for Lavagna's resignation. From then on, Kirchner was able to politicize economic decision making and considerably reduce technical capacity in economic institutions. To do so, Kirchner fired experts and hired economists with fewer technical credentials but policy positions closer to his own preferences. The incumbent's control over economic policy is considerably higher than in the past. Cristina Fernández de Kirchner has largely maintained these trends (Redrado 2010; Noriega 2010).

Although less influential than in Argentina, economic experts in Bolivia also achieved considerable stability in the Ministry of Finance, particularly in the Vice Ministry of Public Finance. This technocracy gained relevance during the early attempt at market reform initiated by MNR President Víctor Paz Estenssoro (1985–1989) and continued by his successor MIR leader Jaime Paz Zamora (1989–1993). However, these technocratic trends matured during the first government of Gonzalo "Goni" Sánchez de Lozada (1993–1997). A politician from a traditional party, the *Movimiento Nacionalista Revolucionario* (MNR), and technocrat himself, Goni had a highly technical style of government and adopted broad reforms in line with IFI's recommendations (World Bank 2000). According to George Gray Molina (2007), this was not a case of a president ceding controls to technocrats but rather a technocratic actor himself fostering technical policy making through groups of experts. This continuity remained through the presidency of former dictator Hugo Banzer (1997–2001), who resigned due to health problems and had his term completed by his vice president Jorge Quiroga (2001–2002).

However, changes in political and economic conditions eventually toppled experts. The 2000s brought strong and sometimes violent protests against the party system, which was strongly associated with market policies and corruption (Mayorga 2002; Gingerich 2010). Evo Morales, leader of the MAS party, was at the forefront of these protests. Huge mobilizations against water privatization (*Guerra del Agua,* April 2000) and gas exports (*Guerra del Gas* 2003) put the government under a state of siege. In 2002, Sánchez de Lozada narrowly won the presidency with 22.5 percent of the vote over Morales who garnered 20.9 percent. Yet the political landscape had changed significantly from Sánchez de Lozada's first term. Conditions were now very unfriendly for technocrats (and traditional political parties in general), with the president besieged by indigenous and civil society anti-neoliberal protests (Madrid 2008).

Violent repression against the protestors led Sánchez de Lozada to step down in 2003. Vice President Carlos Mesa was able to recover some stability, but protests continued and he also resigned. Morales won fresh elections in 2005 with an astounding 54 percent of the vote that empowered him to initiate sweeping changes. This transformation included reversing several market reforms in an effort to build a new pluricultural and socialist state. Technocrats in the Ministry of Finance were associated with the old neoliberal regime and consequently saw their power reduced. Since then, economic policy has more closely followed the priorities of Morales and MAS in their effort to use the state to fulfill the goals of the so-called Bolivian Revolution.[3]

These two cases show the relevance of the two politics of policy factors for technocratic downfall. First, in both countries technocrats' power weakened when crises abruptly reduced the costs politicians faced for firing experts. Severe economic upheaval in Argentina and political turmoil in Bolivia eliminated the popular legitimacy of experts' technical approach, allowing politicians to replace them. In public opinion, orthodox experts in both countries went from being seen as guarantors against instability to culprits of economic mayhem. Thus for politicians, where technocratic stability and independence had once been a significant political asset, it was now perceived as a liability.

In Argentina, this downfall was gradual. Initially, orthodox experts blamed for the mounting economic crisis lost their positions. These experts were considered part of the neoliberal elite that had acted against the public interest in previous years. Nonetheless, this case shows that firing technical teams in the midst of an economic crisis and filling their positions with political loyalists may still be too risky for politicians. Although Duhalde replaced orthodox experts with heterodox experts in line with popular demands, this did not mean decision making in the Ministry became less technical. Lavagna and his team maintained enough autonomy to limit political intrusion and adopt their own policy preferences. It was only when a partial economic recovery, in part led by the commodity boom, reduced the potential cost of firing experts and, as discussed later, Kirchner gained additional leverage in Congress, that the incumbent strongly reduced the technical level in the Ministry of Economics.

In Bolivia, Evo Morales and his party blamed technocrats for exclusionary politics and limiting the use of natural resources for national development. Because the crisis in this case was more political than economic, the risks of politicizing economic policy in order to gain control over resources were low from the start. Rising gas and mining rents provided Morales with sufficient economic resources to conduct his reforms.

The second factor, a balanced constellation of stakeholders, was also crucial in both cases. Although more clearly in Bolivia, both countries experienced the

[3] It is worth noting, however, that this higher political control over economic policy has not meant an irresponsible handling of macroeconomic policy. Economic officials have less power than in the past, but this has not lead to populist macroeconomic policy. In clear contrast to Venezuela's Bolivarian revolution, Bolivia has reversed several market reforms (privatization, social security, and financial reforms, for example) but keeps macroeconomic stability under check.

emergence of a dominant incumbent that severely unbalanced the constellation of stakeholders and contributed to experts' downfall. As dominant actors, incumbents in Argentina and Bolivia were able to politicize technical policy areas. Morales achieved considerable control over Congress as well as over regional and municipal governments, which limited the ability of the political opposition to block his policies. Similarly, Kirchner gradually built a governing coalition that gained considerable control over Congress. The 2005 legislative election gave the president enough power to dismiss Lavagna. Under these conditions, experts' allies had little capacity to protect them. Experts' antagonizing of stakeholders, as happened in Argentina when Cavallo clashed with businesses and IFIs, meant this support was expected to be even lower.

Before concluding these economic cases, it is worth mentioning that the policy consensus factor also played a significant role in Argentina. Here, heterodox economic experts remained prominent during the nineties even though they lost saliency in other Latin American countries. As such, Duhalde could find alternatives to orthodox experts' policies that were still technical. Similarly, Néstor Kirchner and Cristina Fernández de Kirchner both used anti-market rhetoric to justify their appointment of less well-credentialed heterodox experts in the Ministry of Economics.

Although not a case involving economic technocrats, the story of experts in Venezuela's state oil company (PDVSA) further demonstrates how the politics of policy factors contribute to technical experts' downfall. This case shows how the same mechanisms affecting technocrats in state institutions also operate in public enterprises in Latin America. Few institutions had the prestige or technical capacity of PDVSA. Created when oil was nationalized in 1975, PDVSA quickly became an independent and autonomous institution. Even though other areas of the state were part of their clientelistic networks, the two parties in power at the time respected the autonomy of the corporation in charge of administering the oil rent, and PDVSA became known for its professional level (Corrales and Penfold 2011, 76).

In 1998, Hugo Chávez defeated the remnants of Venezuela's two-party system and assumed the presidency. The former army commander quickly gained popularity among citizens who were tired of worsening economic conditions and the corrupt traditional parties that had dominated politics for forty years. During his first years in office, Chávez adopted a series of measures to concentrate power, including a new constitution approved in 1999. Chávez's control over the state and the horizontal institutions meant to check his power was immense, leading to the development of a competitive authoritarian regime in the early 2000s (Levitsky and Way 2010, 178).

Although Chávez repeatedly clashed with PDVSA in an effort to control its resources, the company maintained significant autonomy for several years. Its prestige, institutional strength, and the crucial role of oil rents in Venezuela kept Chávez in check. However, the 2003 oil strike changed this situation. When the strike failed, Chávez retaliated against PDVSA directors and employees that had previously opposed him and took control of the company. Profiting from previous institutional changes designed to centralize authority in PDVSA in order to make the company more efficient, Chávez appointed a politically subordinated manager that largely eliminated the institution's autonomy. Furthermore, a 2001 law limited the involvement of private enterprises in the petroleum sector by mandating the state have at least a 51 percent stake in all new projects. Private companies lost relevance in oil activities (Corrales and Penfold 2011, 77–82). Within a few years, the state enterprise was entirely subordinate to the government and Chávez had full control over oil rents that allowed him to expand his social programs and strengthen his grip over the state until his death in March 2013.

This case confirms how a dominant incumbent can reduce technical capacity with little opposition from stakeholders. The lack of a strong opposition party or other private stakeholders with sufficient clout to protect PDVSA eroded the institution's power. Additionally, in the midst of a commodity boom with oil prices at record levels, Chávez had considerable space to use the oil rents for his political project and reduce the perceived and immediate costs of firing experts. There were few political costs for interfering with this state enterprise but many gains. Everything was aligned against the technocratic management of PDVSA.

In conclusion, these cases show that entrenched technocratic autonomy can decline when events drastically reduce the political costs of firing experts and the constellation of stakeholders becomes unbalanced. Thus my theory of technocratic autonomy is also useful to explain instances when previously autonomous experts lose influence. In the absence of such dramatic events, technical autonomy as found in my economic episodes seems quite stable.

THE LIMITS OF POLITICAL INCENTIVES AND STRUCTURAL THEORIES FOR THE ANALYSIS OF POLICY MAKING

As mentioned in the introduction, the "output" side of politics deserves much more attention than it has received from Latin Americanists because the power of political and economic forces can only be properly weighted through an analysis of policy outcomes (Carpenter 2001, 11; Weyland 2006, 11–14). My analysis supports this position. It points to the limitations of political-incentives and structural theories of policy making and politics more generally in Latin America. As the episodes I discuss show, these approaches attribute too much weight to particular actors while neglecting the importance of others, such as technocrats.

Theories that focus on political incentives to explain policy choices fail to recognize that politicians' interests alone do not determine policy (Weyland 2002b, 64–66). Although the assumption that politicians dictate policy may be accurate where bureaucracies are weak, my study shows that certain state actors, such as technocrats, can restrain politicians and escape their control in ways not compatible with political-incentives theories. Furthermore, some of the episodes I discuss show that politicians are frequently unaware of the content of policy, including those policies that have significant short- and medium-term political repercussions. Certainly, on other occasions, politicians may defeat experts and reduce the technical level of policy areas. However overall, technocrats in the region are much more autonomous and relevant than political-incentives theories recognize.

It is worth mentioning that these results do not only apply to developing states. Many authors see bureaucratic behavior in developed, institutionalized, and more politically stable countries through an political-incentives lens as the product of a calculated delegation by elected politicians: actors within the state are mere agents of politicians due to politicians' control over resources and legislative capabilities. Policy decisions are assumed to correspond to politicians' desires and goals (McCubbins and Schwartz 1984; Chang, De Figueiredo, and Weingast 2000; Huber and Shipan 2002). Some influential studies have challenged these assumptions, showing that politicians, even in highly institutionalized countries, lack such total control over bureaucracies (Moe 1990; 2005; Carpenter 2001). Politicians cannot always ensure agencies' obedience, bureaucrats possess several strategies for escaping such control, and state agencies can advance their goals and contest efforts to reform them. In short, state actors, whether in Latin America or developed democracies, have much more relevance than political-incentives theories usually acknowledge.

Similarly, structural theories that focus on economic forces, external or internal, as the main determinants of policy underestimate the importance of technocrats. On the one hand, external-structural perspectives exaggerate the power of IFIs over local actors. As my study shows, experts are not subordinated to IFIs. While IFIs provide experts with new policy ideas and may even join forces to push forward particular policies, technical policies in Latin America result from the preferences of domestic experts rather than the foreign structural forces IFIs represents. Furthermore, the cases I discuss show how Latin American states use experts to counterbalance the expertise of these external forces in order to defend themselves against their sometimes unwelcome policy recommendations.

Domestic structural perspectives also exaggerate the power of the private sector over technocrats. These theories assume that dominant business interests have the capacity to transform their preferences into policies by capturing parts of the state and making state actors such as technocrats their agents. This may be true for certain agencies lacking the resources to oppose business interests or a balanced constellation of stakeholders that limits business intrusion. But these theories ignore how technocrats have sometimes used their power to transform the structure of an economy against the wishes of dominate business groups, as happened in Colombia and Peru during recent decades.

The assumptions of both structural and political-incentives theories provide an unrealistic and simplistic account of policy making in the region. I concur with Barbara Geddes's conclusion about economic liberalization in the developing world that a structural-economic paradigm based on the "centrality of economic interests" cannot "account for the most important empirical regularities emerging from these and other studies of the politics of economic liberalization" (Geddes 1995, 214). But a similar criticism can be applied to Geddes's own illuminating efforts to build a new paradigm based on "politicians, their interests, and the political circumstances that shape the way in which they pursue their interests" (Geddes 1995, 214). Both paradigms, political and structural, disregard the role of state actors like technocrats and thus cannot account for important patterns in policy making, and politics in general.

In sum, experts deserve to be considered actors in their own right along with politicians, business actors, and IFIs. They initiate policies according to their preferences and often successfully modify their countries' political and economic landscapes. A proper analysis of policy making in Latin America requires incorporating technocrats, weighting their influence in different policy areas, and analyzing their interaction with other political actors.

PRACTICAL LESSONS FOR POLICY MAKERS

This work also has relevant implications for policy makers on the limits and possibilities of various reform strategies and complements other works about policy reform in the region (Grindle and Thomas 1991; Grindle 2004; 2000; 1996). First, my findings point to the limitations of universal reform models that disregard the political and economic context in which state agencies operate. More broadly, these findings point to the limitations of technocratic reform strategies in social policy sectors in general. Second, technical institution building is costly, but the payoffs for legitimacy and sustainability in the long term are considerable.

One lesson of my findings is well known by reform teams: some state policy areas are more difficult to change than are others or, considered from another angle, some reformers are more vulnerable than are others. Reformists in policy areas where the factors identified in my theory are favorable have a greater chance of succeeding, as they enjoy more space to act strategically, are informally protected from political pressures, and have many more allies to support them. On the contrary, technical reform teams in sectors where these conditions are unfavorable face stronger constraints. Assuming the same reform strategy can succeed across all state policy areas may produce goals that are impossible for some reform teams to achieve because of the tighter limits they face.

In fact, my theory makes quite pessimistic predictions about the chances of reforming those policy areas where the first two factors are unfavorable for autonomy. To begin with, even starting reforms in these policy areas is difficult due to a lack of political urgency. Even if reform teams advance new policies, they run a higher risk of reversal. Experts in these policy areas lack powerful allies that can limit political intrusion, and politicians can intervene with few consequences for themselves. Although reform teams may achieve important reforms on paper, the implementation and sustainability of those reforms is considerably more difficult.

For example, my health policy episodes suggest the limitations of World Bank strategies to improve health care by strengthening patients' associations (World Bank 2004; Reuben and Belsky 2006). Recipients of health care services in the region tend to be poor and lacking information. As such, they are not strong enough to counterbalance political interests in the way powerful stakeholders do in economic policy areas. These strategies are certainly right to suggest fostering constituencies that can lobby against reform reversals and for better policy. Creating stakeholders through policy making is an option for reform teams that want to make new policies resilient. Yet these new stakeholders need to be powerful enough to oppose politicians in order to guarantee the stability of new policies and institutions, which may surpass what reforms teams can do (Graham et al. 1999, footnote 26). As other authors have highlighted, a well-organized civil society and political actors on the demand side of policy are crucial to achieving good policy outcomes (Evans 2010, 48–50; O'Donnell 1996, 6).

One obvious lesson for policy makers is that properly understanding these limitations allows them to be more aware of the obstacles they face and more realistic about reform strategies. However, there is a more fundamental lesson: technically trained individuals in government may not suffice to conduct broad and sustainable reforms in several policy areas of Latin American states; politics are also crucial. Without solid political support to overcome barriers, and lacking strong stakeholders demanding better policy and opposing particular interests, thorough reforms are unlikely no matter how competent the reform team is. Strong political coalitions and societal allies that increase the need for better policy outcomes are critical to making technocratic solutions relevant, legitimate, and thus more likely to persist. Although experts are frequently allergic to politics and act to insulate themselves from other actors in order to advance their reforms, they paradoxically seem to need politics in order to succeed in these less likely policy areas.

A second important lesson is that there are significant payoffs for institution building. Chapter 4 shows the long-term benefits of strategies to institutionalize professionalism in economic policy making in Colombia. These economic experts invested considerable resources to support the stability, reproduction, and training of technocrats through building institutions, which have proven crucial to technocratic continuity. Nowadays, experts themselves act as a support group for technocrats in the public sphere.

The contrast with the Peruvian case is revealing. There, economic experts easily achieved control over the MEF and, partly as a result, did not pursue similar strategies to institutionalize technical reproduction in the state and society. In MEF, there are no state-sponsored incentives for training future generations of experts. Technocrats are linked to few think tanks (the *Centro de Investigación de la Universidad del Pacífico*, CIUP, *Grupo de Estudios del Desarrollo*, GRADE, and APOYO) but mostly occupy private sector positions when they leave the state. As such, experts' power is not rooted in civil society or in academic institutions like those the Colombian technocracy built. By lacking similar exit options and a technical environment that values academic achievements and research, Peruvian experts end up more closely associated to, and dependent on, business actors. Consequently, they remain more vulnerable to abrupt political changes, as their proximity to business makes *técnicos* easier to delegitimize.

It seems wise for public policy makers interested in professionalizing their policy areas to invest resources in creating enduring mechanisms that promote stability, reproduction, and continuity, as economic technocrats did in Colombia. The Colombian episode shows how, although costly and demanding, these strategies offer good results for technical continuity in the state over the medium term. In conclusion, this work provides some lessons to policy makers about the challenges, limitations, and possibilities they face when reforming the state.

TECHNOCRATS, DEMOCRACY, AND BEYOND

The early literature on technocrats in the region claimed that experts had a "natural" affinity for authoritarian regimes (O'Donnell 1973; Collier 1979). Experts' insulated policy style and preference for top-down reforms, made them seem the perfect complement for regimes that based their power on the absence of deliberative processes. In the nineties, scholars questioned this argument by stressing that experts were also important actors in democracies (Centeno and Silva 1998; Domínguez 1997). This second group of authors concluded that there was no natural link between experts and a particular regime type. Experts were and always would be important actors in Latin America, whether under dictatorship or democracy.

This book supports this conclusion: experts remain relevant actors in Latin American democracies. *Técnicos*' power was not tied to authoritarian regimes, and they can still exercise significant influence in democracies. In fact, experts' power under democracy is less precarious, because it is not dependent on a single military leader or small junta as it usually was under authoritarian regimes. Their continuity is not questioned when these regimes fall and remains stable across presidential terms. There is no doubt that technocrats have gained stability across the region in democratic times.

Nonetheless, the question about the compatibility of experts' power and democratic politics remains relevant today. As mentioned in the introduction, various authors have warned about the dangers autonomous technocrats present to democracy in Latin America (Centeno 1993; Conaghan and Malloy 1994; Teichman 1997; Centeno and Silva 1998). These authors perceived technocratic influence as limiting democratic politics in the region: citizens may elect governments according to their preferences, but policy outputs are controlled by technocrats (or, in some accounts, the business actors supporting them). In these accounts, experts' policy choices are insulated and respond more to technical preferences than to democratic processes. As such, these choices can produce social costs by disregarding political consequences or pushing forward development projects without evaluating their broader political and social consequences. Furthermore, political bodies lack the technical capacity to monitor experts in economic institutions (Corrales 2011). Some authors even described political regimes in Latin America as "technocratic democracies," meaning that experts, for good or bad, had considerable power over politics and policy making (Centeno and Silva 1998).

My findings speak to these debates about the power of non-elected actors over democratic politicians. In regard to democratic accountability and the influence of experts, my results suggest some serious shortcomings of Latin American democracies. The episodes show that political bodies frequently lack the ability to match technocrats' expertise. As different instances in this book show, meaningful policies are designed and implemented by experts without elected officials fully understanding their content. Furthermore, in the absence of strong democratic stakeholders, such as political parties or civil society associations, economic experts end up being more responsive to stakeholders such as business actors and IFIs. These powerful actors' preferences weigh more than other social demands that may not reach experts but are still very urgent for the population.

This image of a strong technocracy that largely dictates policy making is more accurate in economic policy than in other policy areas where experts lack similar power over elected politicians. Technocrats' independence in the health sector comes from not only their superior expertise but also careful maneuvering to prevent political backlash. Economic technocrats have considerable latitude to act on their own preferences, sometimes disregarding the desires of their democratically elected superiors. Policy is decided within technical policy areas rather than in dialogue with elected representatives. And experts' preferences are frequently not aligned with electoral preferences.

For the last decade, technocratic policy making in Peru has been several steps closer to economic orthodoxy than what presidents promised on the campaign trail. Presidents have bowed to technocracy, and by doing so, have broken many of their campaign promises. As a result, electoral change has not brought policy variation (Vergara 2012). Alan García ran a centrist campaign, but his economic policy remained quite orthodox. More striking was Ollanta Humala's decision to maintain economic continuity after he promised to move toward a less orthodox development model. Part of this economic continuity is explained by the well-entrenched economic technocracy, which business actors and investors perceive as a guarantee of market stability. Although politicians in Colombia usually make campaign promises more in line with what experts support, technocrats frequently limit democratically elected politicians who are aiming to advance demands from their constituencies and generally maintain a firm set of core policies from term to term. As in Peru, Colombian experts have a closer relation with business than with other social sectors. The quality of these democracies where economic experts have amassed considerable power is therefore questionable, as public policies – for good or for bad – do not necessarily conform to the preferences of elected politicians and are more responsive to some interests in society than to others.

But there is another side to this democratic story. Although technocracies may have undemocratic features that reduce democratic quality, experts may be more responsive to citizens' demands than are elected politicians. Elected politicians are frequently unresponsive to citizens' needs because their interest in acquiring patronage resources and advancing particular interests leads them to disregard public preferences. In countries with weak political parties and amateur politicians, such lack of democratic accountability is likely to be significant. As such, politicians also fail the test of democratic accountability. In social policy sectors, for example, politicians pursuing patronage resources and particular interests may create greater challenges in satisfying citizens' demands than do insulated technocrats.

In some respects, then, experts are salutary for a region characterized by patronage politics, regime instability, and weak bureaucracies. As Weber highlighted in regard to bureaucracies, adequate organizational incentives and a lack of concern for electoral politics turn bureaucrats into a positive counterweight against politicians' short-term electoral interests (Weber 1946). This book shows that technocrats can sometimes have similar effects by restraining the patronage-driven impulses of politicians and strengthening institutions in a region where politicians are well known for their irresponsible policy making. Effective professional bureaucracies may be the best solution for ambitious politicians in the region, but in their absence, technocrats act as a "second best" option.

As discussed throughout these chapters, experts are often the ones who initiate reform efforts. Health policy experts have advanced reforms in areas, such as rural health care, that political leaders have largely ignored (Nelson 2004, 31). Similarly, economic technocrats have helped keep inflation under control and maintain tight macroeconomic stability, which has been crucial to legitimizing current democratic regimes. Furthermore, experts' networks are responsible for incorporating locally new development suggestions with some quite positive outcomes. Those who exclusively focus on experts' policy experiments that had poor outcomes should not forget other policies that produced positive outcomes, which then diffused through technical networks. Different works mention policy diffusion through high-ranking state actors such as technocrats as an important mechanism of policy reform in the developing world (Weyland 2002a; Ewig 2010; Haggard and Kaufman 1995). As part of international epistemic communities, technocrats are aware of reform experiences around the world.

Recent works also point to the crucial role experts play in connecting centers of knowledge such as universities and think tanks with the state (Santiso and Whitehead 2006; Tanaka et al. 2011). Technocrats can enrich political debate by bringing in the proposals of other experts outside the state. Further, their proximity to new policy ideas allows technocrats to "import" this knowledge to their countries. In these ways, technocrats provide support for political stability and policy advancement in Latin American democracies.

Of course, these benefits do not reduce the above mentioned democratic tension between insulated technocratic policies and elected politicians. But they highlight that elected politicians are not always responsive to acute problems in their societies and that experts can ameliorate this problem of democratic accountability. A better democratic outcome will undoubtedly require stronger representative institutions capable of matching experts' power. At the same time, higher quality democracies need more responsible politicians who consider both the short- and long-term interests of their entire societies.

These debates will continue in the following decades. As this book shows, technocrats adjust well to regime changes and diverse political and economic conditions. Due to the difficulties of civil service reform, the weakness of regional bureaucracies that makes politicians seek experts' advice and support is not likely to change anytime soon (Grindle 2012). -Nothing indicates that experts will not continue exercising significant power in Latin America for the foreseeable future.

Bibliography

Abel, Christopher (1996). *Ensayos de Historia de la Salud en Colombia, 1920–1990*, Bogotá: Universidad Nacional/IEPRI.

Abusada, Roberto et al. (2000). "La Reforma Incompleta," in Abusada et al. (Eds.), *La Reforma Incompleta: Rescatando los Noventa*, Lima: Centro de Investigación de la Universidad del Pacífico.

Ahumada, Consuelo (1996). *El Modelo Neoliberal y Su Impacto en la Sociedad Colombiana*, Bogotá: El Áncora Editores.

Alvarado Pérez, Betty, and Rony Lenz (2006). "Pro-poor Policies in the Peruvian Public Health Sector," in Daniel Cotlear (Ed.), *New Social Contract for Peru*, Washington, DC: The World Bank.

Amadeo, Edward (Ed.) (1993). *Estabilización y Reforma Estructural en América Latina*, Bogotá: Fedesarrollo/Tercer Mundo Editores.

Ames, Barry (1987). *Political Survival: Politicians and Public Policy in Latin America*, Berkeley: University of California Press.

ANIF (2010). "Emergencia Social en Salud: ¿Acaso Es un Hecho Sobreviniente?" *Informe Semanal*, 1012: February.

(2009). "Universalización de la Salud y Activismo Jurídico," *Informe Semanal*, 993: September.

(2008). "El Costo de la Universalización de la Salud en Colombia," *Informe Semanal*, 947: September.

Arce, Moisés (2006). *Market Reform in Society: Post Crisis Politics and Economic Change in Authoritarian Peru*, University Park: Pennsylvania State University Press.

Arenas, Roberto (1972). "*Texto de la Conferencia Dictada por el Doctor Roberto Arenas Bonilla, Jefe del Departamento Nacional de Planeación, en el Salón Rojo del Hotel Tequendama*," Bogotá: DNP.

Arroyo, Juan (2001). *Gobernabilidad en Salud en el Perú en los Noventa*, Lima: MINSA, Oficina General de Epidemiología.

(2000). *Salud: La Reforma Silenciosa*, Lima: Universidad Cayetano Heredia.

(1999). "La Reforma del Sector Salud en el Perú, 1990–1998: En Busca de Nuevos Modelos de Política Social," in Felipe Portocarrero (Ed.), *Pobreza y Políticas*

Sociales en el Perú: Nuevos Aportes, Lima: Red para el Desarrollo de las Ciencias Sociales.

Babb, Sarah (2005). "Del Nacionalismo al Neoliberalismo: El Ascenso de los Nuevos *Money Doctors* en México," in Jairo Estrada Álvarez (Ed.), *Intelectuales, Tecnócratas, y Reformas Neoliberales en América Latina*, Bogotá: Universidad Nacional de Colombia/Convenio Andrés Bello/Conciencias.

(2001). *Managing Mexico: Economists from Nationalism to Neoliberalism*, Princeton: Princeton University Press.

Banco Interamericano de Desarrollo (1991). *Progreso Económico y Social en América Latina*, Washington, DC: IDB.

Barnett, Michael N., and Martha Finnemore (1999). "The Politics, Power, and Pathologies of International Organizations," *International Organizations*, 53:4 (autumn), 699–732.

Bates, Robert H., and Anne O. Krueger (1993) "Generalizations arising from the Country Studies" in Robert H. Bates and Anne O. Krueger (Eds.) *Political and Economic Interactions in Economic Policy Reform*, Cambridge: Blackwell.

Bell, Daniel (1973). *The Coming of Post-Industrial Society: A Venture in Social Forecasting*, New York: Basic Books.

Berry, Albert (1980). "The National Front and Colombia's Economic Development," in Albert Berry, Ronald G. Hellman, and Mauricio Solaun (Eds.), *Politics of Compromise*, New Brunswick: Transaction Books.

Berry, Albert, and Mauricio Solaun (1980). "Notes toward an Interpretation of the National Front," in Albert Berry, Ronald G. Hellman, and Mauricio Solaun (Eds.), *Politics of Compromise*, New Brunswick: Transaction Books.

Berry, Albert, and Francisco Thoumi (1977). "Import Substitution and Beyond: Colombia," *World Development*, 5:1/2, 89–109.

Bersch, Katherine, Sergio Praça, and Michael Taylor (2012). "An Archipelago of Excellence? Autonomous State Capacity among Brazilian Federal Agencies," unpublished working paper, February 2012.

Biglaiser, Glen (2002). *Guardians of the Nation? Economists, Generals, and Economic Reform in Latin America*, Notre Dame: Notre Dame University Press.

Bill Chávez, Rebeca (2004). *The Rule of Law in Nascent Democracies: Judicial Politics in Argentina*, Stanford: Stanford University Press.

Botero, Rodrigo (2010). "Modernidad, Tecnocracia, y Democracia Liberal," *Coyuntura Económica*, 2010:1, 206–210.

(2005). "Una Nota sobre la Tecnocracia Colombiana," *Coyuntura Económica*, 2005:2, 17–24.

(2004). *El Discreto Encanto de la Social Democracia*, Bogotá: EAFIT.

(1995). "Reflexiones Acerca del Origen y Primeras Actividades de Fedesarrollo," in Héctor Gómez Buendía (Ed.), *Economía y Opinión, 25 Años de Fedesarrollo*, Bogotá: Editor Tercer Mundo Editores, Colciencias.

(1976). *Memorias de Hacienda 1974–1976*. Bogotá: Ministerio de Hacienda.

Boylan, Delia (1998). "Preemptive Strike: Central Bank Reform in Chile's Transition from Authoritarian Rule," *Comparative Politics*, 30:4, 443–462.

Bradbury, Robert Carl (1975). *Planning for Health in Colombia: Confronting Theory and Practice*, Unpublished PhD Book. Ohio State University.

Braun, Miguel et al. (2007). "*Caso de Estudio Colombiano: Fedesarrollo*," CIPPEC.

Brint, Steven (1990). "Rethinking the Policy Influence of Experts: From General Characterizations to Analysis of Variation," *Sociological Forum*, 5:3, 361–385.

Caballero Argáez, Carlos (2008). "Las Etapas de la Planeación en Colombia," *Planeación y Desarrollo*, XXXIX:2, (July–December), 56–65.

Cardoso, Fernando Henrique, and Enzo Faletto (1979). *Dependency and Development in Latin America*, Berkeley and Los Angeles: University of California Press.

Carey, John, and Matthew Shugart (1995). "Incentives to Cultivate a Personal Vote." *Electoral Studies*, 14: 4, 417–439.

Carpenter, Daniel (2001). *The Forging of Bureaucratic Autonomy: Reputations, Networks, and Policy Innovation in Executive Agencies*, Princeton: Princeton University Press.

Centeno, Miguel Ángel (1994). *Democracy within Reason: Technocratic Revolution in Mexico*, University Park: Pennsylvania State University Press.

(1993). "The New Leviathan: The Dynamics and Limits of Technocracy," *Theory and Society*, 22:3, 307–335.

Centeno, Miguel Ángel, and Patricio Silva (1998). "The Politics of Expertise in Latin America: Introduction," in Miguel A. Centeno and Patricio Silva (Eds.), *The Politics of Expertise in Latin America*, New York: Macmillan Press.

Cepeda Ulloa, Fernando (1994). *Dirección Política de la Reforma Económica en Colombia*, Bogotá: FONADE, DNP.

Cepeda Ulloa, Fernando, and Christopher Mitchell (1980). "The Trend toward Technocracy," in Albert Berry, Ronald Hellman, and Mauricio Solaún (Eds.), *Politics of Compromise: Coalition Government in Colombia*, New Brunswick: Transaction Books.

Chang, Kelly H., Rui J. P. De Figueiredo, and Barry Weingast (2000). "Rational Choice Theories of Bureaucratic Control and Performance," in William F. Shugart II and Laura Razzolini (Eds.), *The Elgar Companion to Public Choice*, Cheltenham: Edward Elgar Press.

Chon, Enrique, Vilma Montañez, and Janice Seinfeld (2011). *La Salud que los Peruanos Queremos*, Lima: Centro de Investigación de la Universidad del Pacífico.

Clavijo, Sergio, and Maria Olga Peña (2010). "Reformas al Régimen de Salud: Entre la Emergencia Social y el Déficit Estructural," *ADEGOB*, 5: (September), 4–8.

Collier, David (Ed.) (1979). *The New Authoritarianism in Latin America*, Princeton: Princeton University Press.

Collier, David, and Ruth Berins Collier (1991). *Shaping the Political Arena: Critical Junctures, the Labor Movement, and Regime Dynamics in Latin America*, Notre Dame: University of Notre Dame Press.

Colombia Health Sector Reform Project (1996). *Report on Colombia Health Sector Reform and Proposed Master Implementation Plan*, Boston: Harvard School of Public Health.

Conaghan, Catherine M. (1998). "Stars of the Crisis: The Ascent of the Economists in Peru," in Miguel A. Centeno and Patricio Silva (Eds.), *The Politics of Expertise in Latin America*, New York: Macmillan Press.

Conaghan, Catherine M., and James M. Malloy (1994). *Unsettling Statecraft: Democracy and Neoliberalism in the Central Andes*, Pittsburgh: University of Pittsburgh Press.

Congreso de la República del Perú (2001). Comisión Investigadora sobre los Delitos Económicos y Financieros Cometidos entre 1990–2001. Reporte de Investigación (Agosto–Diciembre 2001). http://www.congreso.gob.pe/comisiones/2002/CIDEF/preliminares/infopre1.pdf.

Corporación para el Fomento de las Investigaciones Económicas (CORP) (1972). *Controversia sobre el Plan de Desarrollo*. Bogotá: Editorial La Oveja Negra.

Corrales, Javier (2011). "Technocratic Policy Making and Parliamentary Accountability in Argentina, 1983–2002," Democracy, Governance, and Human Rights Program Paper N. 13, September 2004.

Corrales, Javier, and Michael Penfold (2011). *Dragon in the Tropics. Hugo Chávez and the Political Economy of Revolution in Venezuela*, Washington: Brookings Institution Press.

Cortázar, Juan Carlos (2006). *La Reforma de la Administración Pública Peruana (1990–1997)*, Lima: Instituto de Políticas Sociales.

Cueto, Marcos (2004). "Origins of Primary Health Care and Selective Primary Health Care," *American Journal of Public Health*, 94:11, 1864–1874.

Dancourt, Óscar et al. (1990). "Una Propuesta de Reforma Monetaria para Acabar con la Hiperinflación," Serie Documentos de Trabajo N. 90, Lima: Departamento de Economía, Pontificia Universidad Católica del Perú.

Dargent, Eduardo (2008). "*Islas de Eficiencia y Reforma del Estado: Ministerios de Economía y Salud, 1990–2008,*" Consorcio de Investigación Económica y Social (CIES), http://www.cies.org.pe/node/1598

Dávila Ladrón de Guevara, Andrés and Natalia Delgado Varela (2002). "La Metamorfosis del Sistema Político Colombiano: ¿Clientelismo de Mercado o Nuevas Formas de Intermediación? in Francisco Gutierrez (Ed.), *¿Degradaciòn o Cambio? Evolución del Sistema Político Colombiano*, Bogotá: Norma.

De Brigard, Alberto (1995). "El Papel de Francisco Ortega en la Independencia de la Banca Central Colombiana," in Roberto Steiner (Comp.), *La Autonomía del Banco de la República: Economía Política de la Reforma*, Bogotá: TM Editores/Fedesarrollo.

Departamento Nacional de Planeación (1974). *La Política de Salud*, Bogotá: DNP/Osprey Impresores

Díaz-Alejandro, Carlos F. (1977). *Foreign Trade Regimes and Economic Development: Colombia* (A Special Conference Series on Foreign Trade Regimes and Economic Development, Volume X), New York and London: National Bureau of Economic Research.

(1973). *El Cambio de una Política de Sustitución de Importaciones a una de Promoción de Exportaciones*, Bogotá: Fedesarrollo.

Dix, Robert (1967). *Colombia: the Political Dimension of Change*, New Haven: Yale University Press.

Domínguez, Jorge (Ed.) (1997). "*Technopols: Freeing Politics and Markets in Latin America in the 1990s*," University Park: Pennsylvania State University Press.

Drake, Paul (1989). *The Money Doctors in the Andes: the Kemmerer Missions 1923–1933*, Durham: Duke University Press.

Durand, Francisco (2006). "El Problema del Fortalecimiento Institucional Empresarial," in John Crabtree (Ed.), *Construir Instituciones: Democracia, Desarrollo, y Desigualdad en el Perú desde 1980*, Lima: Instituto de Estudios Peruanos.

(2005). "Dinámica Política de la Corrupción y Participación Empresarial," in Felipe Portocarrero (Ed.), *El Pacto Infame: La Persistencia de la Corrupción en el Perú*. Lima: Red para el Desarrollo de las Ciencias Sociales en el Perú (IEP, UP, PUCP).

Echebarria, Koldo (2006). "Introducción," in Koldo Echebarria (Ed.), *Informe sobre la Situación del Servicio Civil en América Latina*, Washington, DC: Banco Interamericano de Desarrollo.

Edwards, Sebastián, and Roberto Steiner (2008). *La Revolución Incompleta: Las Reformas de Gaviria*, Bogotá: Norma.

ENAHO (2012). Encuesta Nacional de Hogares. Instituto Nacional de Estadística e Informática. http://www.inei.gob.pe/

Escobar, María Luisa et al. (2009). "Colombia: After a Decade of Health System Reform," in Amanda Glassman et al. (Eds.), *From Few to Many: Ten Years of Health Insurance Expansion in Colombia*, Washington: IDB/Brookings.

Espinoza, Rubén (1998). *Análisis del Proceso de Reforma en el Perú. Informe del Coordinador del Componente de Reestructuración Sectorial del Programa de Fortalecimiento de Servicios de Salud. Noviembre de 1995–Julio de 1998*, Lima: MINSA/PFSS.

Estrada Álvarez, Jairo (2005). "Élites Intelectuales y Producción de Política Económica en Colombia," in Jairo Estrada Álvarez (Ed.), *Intelectuales, Tecnócratas, y Reformas Neoliberales en América Latina*, Bogotá: Universidad Nacional de Colombia/ Convenio Andrés Bello/Conciencias.

(2004). *Construcción del Modelo Neoliberal en Colombia 1970–2004*, Bogotá: Ediciones Aurora.

Evans, Peter (2010). "Constructing the 21st Century Developmental State: Potentialities and Pitfalls," in Omano Edigheji (Ed.), *Constructing a Democratic Developmental State in South Africa: Potentials and Challenges*, Cape Town: HSRC Press, 37–58.

(1992). "The State as Problem and Solution: Predation, Embedded Autonomy, and Structural Change," in Stephan Haggard and Robert R. Kaufman (Eds.), *The Politics of Adjustment*, Princeton: Princeton University Press.

(1979). *Dependent Development*. Princeton: Princeton University Press.

Ewig, Christina (2010). *Second-Wave Neoliberalism: Gender, Race, and Health Sector Reform in Peru*, University Park: Pennsylvania State Press.

(2004). "Piecemeal but Innovative: Health Sector Reform in Peru," in Robert R. Kaufman and Joan M. Nelson (Eds.), *Crucial Needs, Weak Incentives: Social Sector Reform, Democratization, and Globalization in Latin America*. Baltimore: Woodrow Wilson Center and Johns Hopkins University Press, 2004.

(1999). "Democracia Diferida: Un Análisis del Proceso de Reforma del Sector Salud," in Felipe Portocarrero (Ed.), *Pobreza y Políticas Sociales en el Perú: Nuevos Aportes*, Lima: Red para el Desarrollo de las Ciencias Sociales.

Falleti, Tulia (2010). *Decentralization and Subnational Politics in Latin America*, Cambridge: Cambridge University Press.

Fedesarrollo (2010). "El Sistema de Salud, la Equidad, las Finanzas Públicas y la Emergencia Social: ¿Vamos por el Camino Correcto?" *Tendencia Económica*, N. 93.

(2009). *Informe: Descentralización, el Financiamiento de la Salud y la Educación y los Departamentos*, Bogotá: Fedesarrollo, Federación Nacional de Departamentos.

Flórez Enciso, Luis Bernardo (2010). "Colombia: Economics, Economic Policy, and Economists," in Veronica Montecinos and John Markoff (Eds.), *Economists in the Americas*, Cheltenham: Edward Elgar.

Francke, Pedro (2001). "Lineamientos en Políticas de Salud (2001–2006)," in Pedro Francke (Ed.) *Políticas de Salud 2001–2006*, Lima: Consorcio de Investigación Económica y Social, 21–35.

Franco, Rolando (1996). "Los Paradigmas de la Política Social en América Latina," Revista de la CEPAL, 58, 9–22.

Frank, André Gunder (1966). "The Development of Underdevelopment." *Monthly Review*, 17 (September), 17–31.

Frieden, Jeffry (1985). *Debt, Development, and Democracy: Modern Political Economy and Latin America, 1965–1985*, Princeton: Princeton University Press.

Freidson, Eliot (1984). "Are the Professions Necessary?" in Thomas Haskell (Ed.), *The Authority of Experts: Studies in History and Theory*, Bloomington, IN: Indiana University Press.

Gaviria, Alejandro, Carlos Medina, Carolina Mejía, David McKenzie, and Rodrigo Soares (2006). "Assessing Health Reform in Colombia: From Theory to Practice," *Economía*, 7:1 (Fall), 29–72.

Geddes, Barbara (1995). "The Politics of Economic Liberalization," (Review), *Latin American Research Review*, 30:2, 195–214.

(1994). *Politician's Dilemma: Building State Capacity in Latin America*, Berkeley: University of California Press.

(1990). "Building 'State' Autonomy in Brazil, 1930–1964," *Comparative Politics*, 22:2, 217–235.

Ghezzi, Piero, and José Gallardo (2013). *¿Qué se Puede hacer con el Perú? Ideas para Sostener el Crecimiento Económico en el Largo Plazo*, Lima: Universidad del Pacífico, Fondo Editorial de la Pontificia Universidad Católica del Perú.

Giedion, Ursula, Luis Morales, and Olga Acosta (2000). "Desarrollo Institucional del Sector Salud en Bogotá," *Coyuntura Social*, 23, 57–81.

Gingerich, Daniel (2010). "Bolivia: Traditional Parties, the State, and the Toll of Corruption," in Stephen Morris and Charles Blake (Eds.), *Corruption and Politics in Latin America: National and Regional Dynamics*, London: Lynne Rienner Publishers.

Ginsburg, Tom (2003). *Judicial Review in New Democracies: Constitutional Courts in East Asia*, Cambridge: Cambridge University Press.

Glassman, Amanda et al. (Eds.) (2009). *From Few to Many, Ten Years of Health Insurance Expansion in Colombia*, Washington: IDB/Brookings.

González-Rossetti, Alejandra, and Patricia Ramírez (2000). *Enhancing the Feasibility of Health Reform: the Colombian Case*, Latin America and the Caribbean Health Sector Reform Initiative N. 39, Harvard School of Public Health.

Graham, Carol, Merilee Grindle, Eduardo Lora, and Jessica Seddon (1999). *Improving the Odds: Political Strategies for Institutional Reform in Latin America*, Washington: Inter-American Development Bank.

Gray Molina, George (2007). "El Reto Posneoliberal de Bolivia," *Nueva Sociedad*, N209, May–June 2007, 118–129.

Gray Molina, George, Ernesto Yépez de Rada, and Ernesto Yañez (1999). *La Economía Política de las Reformas Institucionales en Bolivia*, Washington: IADB.

Grindle, Merilee (2012). *Jobs for the Boys*, Cambridge and London: Harvard University Press.

(2004). *Despite the Odds: The Contentious Politics of Education Reforms*, Princeton and Oxford: Princeton University Press.

(2000). *Audacious Reforms: Institutional Invention and Democracy in Latin America*, Baltimore: Johns Hopkins University Press.

(1996). *Challenging the State: Crisis and Innovation in Latin America and Africa*, London: Cambridge University Press.

(1977). "Power, Expertise, and the 'Técnico': Suggestions from a Mexican Case Study," *The Journal of Politics* 39:2,399–426.

Grindle, Merilee, and John W. Thomas (1991). *Public Choice and Policy Change: The Political Economy of Reform in Developing Countries*, Baltimore and London: Johns Hopkins University Press.

Grupo de Economía de la Salud (2007). "Reforma a la Regulación en Salud en Colombia: Del Consejo Nacional a la Comisión de Regulación," *Observatorio de la Seguridad Social*, Universidad de Antioquia, 6:16, 1–12.

Gutiérrez Sanín, Francisco (2007). *Lo que el Viento se Llevó: Los Partidos Políticos y la Democracia en Colombia (1958–2006)*, Bogotá: Norma.

(1997). "La Experiencia de Fedesarrollo," in Hernando Gomez and Hernán Jaramillo (Eds.) *27 Formas de Hacer Ciencia en América Latina*, Bogota: Tercer Mundo, Colciencias.

Gutiérrez Sanín, Francisco, and Andrés Dávila (2000). "Paleontólogos o Politólogos: ¿Qué Podemos Decir Hoy sobre los Dionosurios?" in *Revista de Estudios Sociales*, 6: (May), 39–50.

Haas, Peter M. (1992). "Introduction: Epistemic Communities and International Policy Coordination," *International Organization*, 46:1, 1–35.

Haggard, Stephan, and Robert Kaufman (1995). *The Political Economy of Democratic Transitions*, Princeton: Princeton University Press.

Hagopian, Frances (1994). "Traditional Politics against State Transformation in Brazil," in Joel Migdal, Atul Kohli, and Vivienne Shue (Eds.), *State Power and Social Forces*, Cambridge: Cambridge University Press.

Hall, Peter (1993). "Policy Paradigms, Social Learning, and the State. The Case of Economic Policy Making in Britain," *Comparative Politics*, 25:(April), 275–296.

Halliday, Terence (1987). *Beyond Monopoly: Lawyers, State Crisis, and Professional Empowerment*, Chicago: University of Chicago Press.

Hartlyn, Jonathan (1988). *The Politics of Coalitional Rule in Colombia*, Cambridge: Cambridge University Press.

Haskell, Thomas (1984). "Introduction," in Thomas Haskell (Ed.), *The Authority of Experts: Studies in History and Theory*, Bloomington, IN: Indiana University Press.

Hershberg, Eric (2006). "Technocrats, Citizens, and Second Generation Reforms: Colombia's Andean Malaise," in Eric Hershberg and Paul Drake (Eds.), *State and Society in Conflict: Comparative Perspectives on the Andean Crisis*, Pittsburgh: University of Pittsburgh Press.

Hildebrandt, Mary, and Merilee Grindle (1997). "Building Sustainable Capacity in the Public Sector. What Can Be Done?" in Merilee Grindle (Ed.), *Getting Good*

Government: Capacity Building in the Public Sectors of Developing Countries, Cambridge, MA: Harvard Institute for International Development.

Hirschman, Albert (1970). *Exit, Voice, Loyalty: Responses to Decline in Firms, Organizations, and States*, Cambridge, MA: Harvard University Press.

Hommes, Rudolf (1995). "El Proyecto del Gobierno y Su Evolución durante la Constituyente," in Roberto Steiner (Comp.), *La Autonomía del Banco de la República: Economía Política de la Reforma*. Bogotá: TM Editores/Fedesarrollo.

(1988). "La Insoportable Oscuridad del Emisor," Is Estrategia Económica y Financiera, September No. 25, pp. 27–31.

Huber, John D., and Charles R. Shipan (2002). *Deliberate Discretion? The Institutional Foundations of Bureaucratic Autonomy*, New York: Cambridge University Press.

Huber, Evelyne (1995). "Assessments of State Strength," in Peter Smith (Ed.), *Latin America in Comparative Perspective: New Approaches to Methods and Analysis*, Boulder, San Francisco, Oxford: Westview Press.

IOP (2013). Informe de Estudio: Conflictos Sociales y Representación Política, Lima: Instituto de Opinión Pública-PUCP.

Izquierda Unida (1990). *Plan de Gobierno de Izquierda Unida (1990–1995)*, Lima: no press.

(1985) *Plan de Gobierno (1985–1990)*, Lima: no press.

Jaramillo, Miguel, and Sandro Parodi (2004). "El Seguro Escolar Gratuito y el Seguro Materno Infantil: Análisis de Su Incidencia e Impacto sobre el Acceso a los Servicios de Salud y sobre la Equidad en el Acceso," Documento de trabajo No. 46. Lima: GRADE.

Johnson, Chalmers A. (1995). *Japan: Who Governs? The Rise of the Developmental State*, New York: W. W. Norton.

Johnson, Matthew (forthcoming). "The Political Logic of Renters Insurance," Unpublished article.

Juárez, Carlos E. (1995). *The Political Economy of Economic Policy Reform in Colombia: Technocratic Bureaucracy and Business-Government Relations, 1966–1992*. Unpublished PhD Book, University of California, Los Angeles.

(1993a). "Trade and Development Policies in Colombia: Export Promotion and Outward Orientation, 1967–1992," *Studies in Comparative Political Development*, 28:3, 67–97.

(1993b). "Economía Política y Apertura Económica: Desarrollo Económico y Comercio Exterior en Colombia, 1967–1991." *Monograph Series N. 33 Departamento de Administración*, Bogotá: Universidad de los Andes.

Kahler, Miles (1992). "External Influence, Conditionality, and the Politics of Adjustment," in Stephan Haggard and Robert R. Kaufman (Eds.), *The Politics of Adjustment*, Princeton: Princeton University Press.

Kaufmann, Daniel, Massino Mastruzzi, and Diego Zavaleta (2003). "Sustained Macroeconomic Reforms, Tepid Growth: A Governance Puzzle in Bolivia?" in: Dani Rodrik (Ed.), *In Search of Prosperity: Analytical Narratives on Economic Growth*, Princeton: Princeton University Press.

Kaufman, Robert R., and Joan M. Nelson (Eds.) (2004). *Crucial Needs, Weak Incentives: Social Sector Reform, Democratization, and Globalization in Latin America*. Baltimore: Woodrow Wilson Center and Johns Hopkins University Press.

Kitschelt, Herbert and Steven I. Wilkinson (2007). "Citizen-Politician Linkages: An Introducion", in: Herbert Kitschelt and Steven I. Wilkinson (Eds.), *Patrons, Clients and Policies*, Cambridge: Cambridge University Press.

Kline, Harvey (1983). *Colombia: Portrait of Unity and Diversity*, Boulder, Colorado: Westview Press.

Koivumaeki, Riita Ilona (2010). "Business, Economic Experts, and Conservative Party Building in El Salvador," *Journal of Politics in Latin America*, 2:1, 79–106.

Krasner, Stephen D. (1984). "Approaches to the State: Alternative Conceptions and Historical Dynamics," *Comparative Politics*, 16:2, 223–246.

Leal Buitrago, Francisco, and Andrés Dávila (1990). *Clientelismo: El Sistema Político y Su Expresión Regional*, Bogotá: Tercer Mundo Editores, IEPRI, Universidad Nacional de Colombia.

Leiteritz, Ralf (2010). *Sustaining Open Capital Accounts. International Norms and Domestic Institutions: A Comparison between Perú and Colombia*, Unpublished PhD Book, London School of Economics.

Levitsky, Steven, and María Victoria Murillo (2009). "Variation in Institutional Strength," *The Annual Review of Political Science*, 11, 115–133.

Levitsky, Steven, and Lucan Way (2010). *Competitive Authoritarianism*, Cambridge: Cambridge University Press.

Leys, Collin (1996). *The Rise and Fall of Development Theory*, Oxford: James Curry.

Lleras Restrepo, Carlos (1965). "Los Programas de la Gran Transformación," in *Discursos*, Bogotá: no press.

Londoño, Juan Luis, and Julio Frenk (1997). "Structured Pluralism: Toward an Innovative Model for Health System Reform in Latin America," *Health Policy*, 41:1, 1–36.

Lora, Eduardo (Ed.) (1991). *Apertura y Crecimiento: El Reto de los Noventa*, Bogotá: Fedesarrollo/Tercer Mundo.

Lora, Eduardo, and Cristina Lanzetta (Comps.) (1995). *El Salto Social en Discusión*, Bogotá: Fedesarrollo.

Lora, Eduardo, and Carlos Scartascini (Eds.) (2009). *Consecuencias Imprevistas de la Constitución de 1991: La Influencia de la Política en las Políticas Económicas*, Bogotá, Alfaomega/Fedesarrollo.

Lowenthal, Abraham (Ed.) (1975). *The Peruvian Experiment: Continuity and Change under Military Rule*, Princeton: Princeton University Press.

Madrid, Raúl (2008). "The Rise of Ethnopopulism in Latin America," *World Politics*, 60: 3, 475–508.

Magaloni, Beatriz (2006). *Voting for Autocracy: Hegemonic Party Survival and Its Demise in Mexico*, Cambridge: Cambridge University Press.

(2003). "Authoritarianism, Democracy, and the Supreme Court: Horizontal Exchange and the Rule of Law in Mexico," in Scott Mainwaring and Christopher Welna (Eds.), *Democratic Accountability in Latin America*, Oxford: Oxford University Press, 267–305.

Mahoney, James (2001). *The Legacies of Liberalism: Path Dependence and Political Regimes in Central America*, Baltimore: Johns Hopkins University Press.

Mahoney, James, and Diana Rodríguez-Franco (forthcoming). "Dependency Theory," Prepared for inclusion in the *Handbook on the Politics of Development*, edited by Carol Lancaster and Nicolas van de Walle.

Malloy, James (1992). "El Problema de la Gobernabilidad en Bolivia, Perú, y Ecuador" in René Mayorga (Ed.) *Democracia y Gobernabilidad: América Latina*, Caracas: Nueva Sociedad.

Malpica, Carlos (1985). *La Alternativa de la Izquierda: Presupuesto 1985*, Lima: Servicios Populares.

Mann, Michael (1984). "The Autonomous Power of the State: Its Origins, Mechanisms, and Results," *Archives Européennes de Sociologie*, 25, 185–213.

Mares, David R. (1993). "State Leadership in Economic Policy: A Collective Action Framework with a Colombian Case," *Comparative Politics*, 24:4, 455–473.

Mauceri, Philip (1995). "State Reform, Coalitions, and the Neoliberal Autogolpe in Peru," *Latin American Research Review*, 30:1, 7–37.

Mayorga, Fernando (2007). *Encrucijadas: Ensayos sobre Democracia y Reforma Estatal en Bolivia*. La Paz: Editorial Gente Común, CESU-UMSS.

(2002). "Sistema de Partidos y Neopopulismo en Bolivia," in Marcelo Cavarozzi and Juan Manuel Abal (Eds.), *El Asedio a la Política. Los Partidos Latinoamericanos en la Era Neoliberal*, Rosario: Homo Sapiens, 2002, 187–210.

McCubbins, Mathew D., and Thomas Schwartz (1984). "Congressional Oversight Overlooked: Policy Patrol vs. Fire Alarms," *American Journal of Political Science*, 28, 165–179.

McGuire, James W. (2010). *Wealth, Health, and Democracy in East Asia and Latin America*, Cambridge: Cambridge University Press.

(2001). "Social Policy and Mortality Decline in East Asia and Latin America," *World Development*, 29:10, 1673–1697.

MEF (2006). *La Reforma Constitucional sobre Pensiones: Descripción de la Buena Práctica*. Unpublished draft document.

(2001). "Transparencia Económica y Fiscal." Lima: Ministerio de Economía y Finanzas.

Meisel Roca, Adolfo (1997). "¿Por Qué No Hay Hiperinflación en Colombia?" *Ensayos de Política Económica*, No. 32, 97–137.

Mesa-Lago, Carmelo (1996). *La Reforma del Sistema de Salud en Perú: Evaluación de Documentos y Recomendaciones. Consultoría Internacional con el Programa de Fortalecimiento de Servicios de Salud del MINSA. Lima, 14–20 de Julio*, Lima: MINSA/PFSS.

Meynaud, Jean (1969). *Technocracy*, New York: Free Press.

Miller, Gary, and Terry Moe (1983). "Bureaucrats, Legislators, and the Size of Government," *American Political Science Review*, 77: (June), 297–322.

Ministerio de Salud (1994). "*Recuento del Diseño y Trámite de la Reforma en Seguridad Social en Salud*," Bogotá: Ministerio de Salud.

Ministerio de Salud, DNP (1990). *La Salud en Colombia. Insumo para la Ley 10 de 1990*. Bogotá: DNP/Ministerio de Salud.

MINSA (2001). *Memoria Ministerial: La Recuperación de la Democracia. 25 de Noviembre 2000–26 de Julio 2001*. Lima: Ministerio de Salud.

(1996). *El Desafío del Cambio del Milenio: Un Sector Salud con Equidad, Eficiencia, y Calidad. Lineamientos de Política de Salud 1995–2000*. Lima: MINSA.

(1990a) *Políticas y Plan de Salud, 90–95*, Lima: MINSA.

(1990b). *Sistema Nacional Regionalizado Integrado en Salud*, Lima: Ministerio de Salud.

Moe, Terry M. (2005). "Power and Political Institutions," *Perspectives on Politics*, 3:2, 215–233.

(1990). "Political Institutions: The Neglected Side of the Story," *Journal of Law, Economics, and Organization*, 6, 213–254.

(1984). "The New Economics of Organization," *American Journal of Political Science*, 28 (November), 739–777.

Montecinos, Verónica (2005). "Nota sobre la Evolución e Influencia de la Tecnocracia Económica en Chile," in Jairo Estrada Álvarez (Ed.), *Intelectuales, Tecnócratas, y Reformas Neoliberales en América Latina*, Bogotá: Universidad Nacional de Colombia/Convenio Andrés Bello/Conciencias.

(1998). "Economists and Party Politics: Chilean Democracy in the Era of Markets," in Miguel A. Centeno and Patricio Silva (Eds.), *The Politics of Expertise in Latin America*, New York: Macmillan Press.

Montecinos, Verónica, John Markoff, and María José Álvarez-Rivadulla (2010). "Economists in the Americas: Convergence, Divergence, and Connection," in Verónica Montecinos and John Markoff (Eds.) *Economists in the Americas*, Cheltenham: Edward Elgar Publishing.

Montenegro, Armando (2010). "Presentación del Libro "El Bejuco de Tazán y Otras Disgresiones Tecnocráticas," *Coyuntura Económica*, XL:1, 215–218.

Naranjo Velandia, Durfari, Jairo Humberto Restrepo Zea, and Sandra Rodríguez Acosta (2002). "Organismos Reguladores del Sistema de Salud Colombiano: Conformación, Funcionamiento y Responsabilidades," Universidad de Antioquia. Centro de Investigaciones Económicas. Online at http://mpra.ub.uni-muenchen.de/857/

Nelkin, Dorothy (1975). "The Political Impact of Technical Expertise," *Social Studies of Science*, 5: (February), 35–54.

Nelson, Joan (2004). "The Politics of Health Sector Reform: Cross National Comparisons," in Robert R. Kaufman and Joan M. Nelson (Eds.), *Crucial Needs, Weak Incentives: Social Sector Reform, Democratization, and Globalization in Latin America*. Baltimore: Woodrow Wilson Center and Johns Hopkins University Press, 2004, 23–64.

(1999). *Reforming Health and Education: The World Bank, the IDB, and Complex Institutional Change*, Baltimore: Johns Hopkins University Press.

Noriega, Gustavo (2010). *Indec: Historia Íntima de una Estafa*, Buenos Aires: Sudamericana.

North, Douglass (1995). "New Institutional Economics and Third World Development," in Harris, John, Janet Hunter, and Colin Lewis (Eds.), *The New Institutional Economics and Third World Development*, London and New York: Routledge.

(1990). *Institutions, Institutional Change, and Economic Performance*, New York: Cambridge University Press.

O'Brien, Philip J., and Jaqueline Rodick (1983). *Chile, the Pinochet Decade: The Rise and Fall of the Chicago Boys*, London: Latin American Bureau.

Ocampo, José Antonio, and Eduardo Sarmiento (Eds.) (1987). *¿Hacia un Nuevo Modelo de Desarrollo?* Bogotá: Fedesarrollo/Tercer Mundo Editores.

O'Donnell, Guillermo (1996). "Poverty and Inequality in Latin America: Some Political Reflections," Working Paper N. 225. Kellogg Institute for International Studies, University of Notre Dame.

(1973). *Modernization and Bureaucratic Authoritarianism*, Berkeley: Institute of International Studies, University of California Press.

Ordóñez, Antonio (1991). "La Salud: Una Prioridad en el Gobierno de Transformación Nacional," in *25 Años del Gobierno de Transformación Nacional*. Bogota: Nueva Frontera, 47–56.

(1968). *Bienestar Social en Colombia: Ponencia Presentada por Colombia a la Conferencia Internacional de Ministros de los Servicios de Bienestar Social*, Nueva York 1968, Bogotá: Ministerio de Salud.

Ortiz de Zevallos, Gabriel, Hugo Eyzaguirre, Rosa María Palacios, and Pierina Polarollo (1999). *La Economía Política de las Reformas Institucionales en el Perú: Los Casos de Educación, Salud, y Pensiones*. Working paper R-3948. Washington: Inter-American Development Bank.

Palacios, Marco (2001). "Saber Es Poder: El Caso de los Economistas Colombianos," in *De Populistas, Mandarines, y Violencias*, Bogotá: Editorial Planeta Colombiana.

Panizza, Francisco (2013). "Everybody Out: We Are Fantastic," in Francisco Panizza and George Philip (Eds.), *The Politics of Financial Crises in Argentina and Uruguay 2001–2003*, London: Routledge, 27–46.

Pastor, Manuel, and Carol Wise (1992). "From Orthodoxy to Heterodoxy and Back," *Latin American Research Review*, 27:2, 83–117.

Payne, James L. (1968). *Patterns of Conflict in Colombia*, New Haven: Yale University Press.

Petrera, Margarita, and Janice Seinfeld (2007). *Repensando la Salud en el Perú*, Lima: CIUP.

Pierson, Paul (2000). "*Increasing Returns, Path Dependence, and the Study of Politics*," *American Political Science Review*, 94:2, 251–267.

(1994). *Dismantling the Welfare State: Reagan, Thatcher, and the Politics of Retrenchment*, New York: Cambridge University Press.

Polarollo, Pierina (2000). "La Reforma del Sector Salud," in Abusada et al. (Eds.), *La Reforma Incompleta*, Lima: Centro de Investigación de la Universidad del Pacífico.

Putnam, Robert D. (1977). "Elite Transformation in Advanced Societies: An Empirical Assessment of the Theory of Technocracy," *Comparative Political Studies*, 10:3 (October), 383–412.

Ramírez, Patricia (2004). "A Sweeping Health Reform: the Quest for Unification, Coverage, and Efficiency," in Robert R. Kaufman and Joan M. Nelson (Eds.), *Crucial Needs, Weak Incentives: Social Sector Reform, Democratization, and Globalization in Latin America*. Baltimore: Woodrow Wilson Center and Johns Hopkins University Press, 2004.

Redondo, Hernán, and Fernando Guzmán (1999). *La Reforma de la Salud y la Seguridad Social en Colombia: El Desastre de un Modelo Económico*. Medellín: Biblioteca Jurídica Dike.

Redrado, Martín (2010). *Sin Reservas: Un Límite al Poder Absoluto*, Buenos Aires, Planeta.

Reina, Mauricio, and Sara Zuluaga (2006). "Safeguards and Antidumping Duties in Colombia," in: J. Michael Finger and Julio J. Nogués (Eds.), *Safeguards and Antidumping in Latin American Trade Liberalization: Fighting Fire with Fire*, Washington, DC: World Bank and Palgrave McMillan.

Rentería, Carolina (2008). "Cincuenta Años del DNP: Historia de una Construcción Colegiada," *Planeación y Desarrollo*, XXXIX:2, 84–99.
Reuben, William, and Leah Belsky (2006). "Voice in the Accountability of Social Policy," in Daniel Cotlear (Ed.), *New Social Contract for Peru*, Washington: World Bank.
Reyna, Carlos, and Antonio Zapata (1991). *Crónica sobre el Cólera en el Perú*. Lima: Desco.
Riesman, David (1950) *The Lonely Crows*, New Haven and London: Yale University Press
Rios Figueroa, Julio (2007). "Fragmentation of Power and the Emergence of an Effective Judiciary in Mexico, 1994–2002," *Latin American Politics and Society*, 49:1, 31–57.
Rustow, Dankwart (1970). "Transitions to Democracy: Toward a Dynamic Model," *Comparative Politics*, 2:3, 337–363.
Sandilands, Roger (1990). *The Life and Political Economy of Lauchlin Currie: New Dealer, Presidential Adviser, and Development Economist*, Durham: Duke University Press.
Santiso, Javier, and Laurence Whitehead (2006). "Ulysses, the Sirens, and the Art of Navigation: Political and Technical Rationality in Latin America," Working Paper N. 256. OECD Development Center.
Sarfatti Larson, Magali (1984). "The Production of Expertise and the Constitution of Expert Power," in Thomas Haskell (Ed.), *The Authority of Experts: Studies in History and Theory*, Bloomington, IN: Indiana University Press.
Schaar, John (1984). "Legitimacy in the Modern State," in William Connolly (Ed.), *Legitimacy and the State*, Oxford: Basil Blackwell.
Schamis, Héctor (1999). "Distributional Coalitions and the Politics of Economic Reform in Latin America," *World Politics*, 51:2, 236–268.
Schmidt, Steffen W. (1974). "Bureaucrats as Modernizing Brokers? Clientelism in Colombia," *Comparative Politics*, 6:3, 425–450.
Schneider, Ben Ross (2004). *Business Politics and the State in Twentieth Century Latin America*, Cambridge: Cambridge University Press.
(1998). "The Material Basis of Technocracy: Investor Confidence and Neoliberalism in Latin America," in Miguel A. Centeno and Patricio Silva (Eds.), *The Politics of Expertise in Latin America*, New York: Macmillan Press.
(1993). "The Career Connection: A Comparative Analysis of Bureaucratic Preferences and Insulation," *Comparative Politics*, 25 (April), 331–350.
Shugart, Matthew and John Carey (1992). *Presidents and Assemblies*, Cambridge: Cambridge University Press.
Silva, Eduardo (1993). "Capitalist Coalitions, the State, and Neoliberal Economic Restructuring: Chile, 1973–1978," *World Politics*, 45:4, 526–559.
Silva, Patricio (2008). *In the Name of Reason: Technocrats and Politics in Chile*, University Park: Pennsylvania State University Press.
(1991). "Technocrats and Politics in Chile: From the Chicago Boys to the Cieplan Monks," *Journal of Latin American Studies*, 23:2, 385–410.
Skocpol, Theda (1985). "Bringing the State Back In: Strategies of Analysis in Current Research," in Peter Evans, Dietrich Rueschemeyer, and Theda Skocpol (Eds.), *Bringing the State Back In*, Cambridge: Cambridge University Press.

Stallings, Barbara (1992). "International Influence on Economic Policy: Debt, Stabilization, and Structural Reform," in Stephan Haggard and Robert R. Kaufman (Eds.), *The Politics of Adjustment in the Third World*, Princeton: Princeton University Press.

Stepan, Alfred (1978). *State and Society: Peru in Comparative Perspective*, Princeton: Princeton University Press.

Stokes, Susan (2001). *Mandates and Democracy: Neoliberalism by Surprise in Latin America*, Cambridge and New York: Cambridge University Press.

Suleiman, Ezra (1974). *Politics, Power, and Bureaucracy in France*, Princeton: Princeton University Press.

Tanaka, Martín (2005). *Democracia sin Partidos: Perú 2000–2005*, Lima: IEP.

Tanaka, Martín, Rodrigo Barrenechea, and Jorge Morel (2011). "La Relación entre Investigación y Políticas Públicas en América Latina: Un Análisis Exploratorio," in Norma Correa and Enrique Mendizabal (Eds.), *Vínculos entre Conocimiento y Política: El Rol de la Investigación en el Debate Público en América Latina*. Lima: CIES/Universidad del Pacífico, 2011, 35–82.

Taylor, Matthew (2009). "Institutional Development through Policy Making: A Case Study of the Brazilian Central Bank," *World Politics*, 61:3, 487–515.

Teichman, Judith (2004). "Merging the Modern and the Traditional: Market Reform in Chile and Argentina," *Comparative Politics*, 37:1, 23–40.

(2002). "Private Sector Power and Market Reform: Exploring the Domestic Origins of Argentina's Meltdown and México's Policy Failures," *Third World Quarterly*, 22:3, 491–512.

(1997). "Mexico and Argentina: Economic Reform and Technocratic Decision Making," *Studies in Comparative International Development*, 32:1, 31–55.

Tejada de Rivero, David (1985). "*Conference: Atención Primaria de la Salud y Ampliación de la Cobertura*," in Laura Morales (Ed.), *Políticas de Salud para el Perú: Medidas de Urgencia*, Lima: Friedirich Ebert.

Thorp, Rosemary (1991). *Economic Management and Economic Development in Peru and Colombia*, Pittsburgh: University of Pittsburgh Press.

Thorp, Rosemary, and Francisco Durand (1997). "A Historical View of Business-State Relations: Colombia, Peru, and Venezuela Compared," in Ben Ross Schneider and Sylvia Mansfield (Eds.), *Business and the State in Developing Countries*, Ithaca and London: Cornell University Press.

Tocqueville, Alexis de (2000). *Democracy in America*, Translated and Edited by Harvey Mansfield and Delba Winthrop, Chicago and London: University of Chicago Press.

Trimberger, Ellen Kay (1978). *Revolution from Above: Military Bureaucrats and Development in Japan, Turkey, Egypt, and Peru*, New Brunswick, NJ: Transaction Books.

Ugarte, Mayén (2000) "La Reforma del Estado: Alcances y Perspectivas in Abusada et al. (Eds.), La Reforma Incompleta: Rescatando los Noventa, Lima: Centro de Investigación de la Universidad del Pacífico, 13–61.

Ugarte, Oscar, and Jose Antonio Monje (1999). "Equidad y Reformas en el Sector Salud," in Felipe Portocarrero (Ed.), *Pobreza y Políticas Sociales en el Perú: Nuevos Aportes*. Lima: Red para el Desarrollo de las Ciencias Sociales.

Universidad del Pacífico (1979). *Perú 1980, Planes de Gobierno*, Lima: Universidad del Pacífico.

Uribe, Juan Pablo (2004). "The Influence of Foreign Models on Health Reform in Colombia," in Kurt Weyland (Ed.), *Learning from Foreign Models in Latin*

American Policy Reform, Washington and Baltimore: Woodrow Wilson Center Press and Johns Hopkins University Press.

Urrutia, Miguel (2008). "La Función de la Planeación en Colombia," *Planeación y Desarrollo*, XXXIX:2, (July–December), 66–75.

(1991). "On the Absence of Economic Populism in Colombia," in Rudiger Dornbusch and Sebastian Edwards (Eds.), *The Macroeconomics of Populism in Latin America*, Chicago and London: University of Chicago Press.

(1983). *Gremios, Política Económica, y Democracia*, Bogotá: Fondo Cultural Cafetero.

USAID (1974). *Misión USAID Colombia. Un Análisis del Sector Colombiano de Salud Pública*, Bogotá: USAID.

Van Dijck, Pitou (1998). "The World Bank and the Transformation of Latin American Society," in Miguel A. Centeno and Patricio Silva (Eds.), *The Politics of Expertise in Latin America*. New York: Macmillan Press.

Vergara, Alberto (2012). "Alternancia sin Alternativa: ¿Un Año de Humala o Veinte Años de un Sistema?" *Revista Argumentos*, 6:3 (July). <http://revistargumentos.org.pe/alternancia_sin_alternativa.html>

Vidal, Carlos (1994a). "7 Meses y 7 Días Como Ministro de Salud del Perú," in Carlos Vidal, *Apuntes de una Vida Dedicada a la Gente*, Tomo II, Lima: Sinco Editores, 307–341.

(1994b). "Una Aproximación Diferente al Ministerio de Salud," in Carlos Vidal, *Apuntes de una Vida Dedicada a la Gente*, Tomo II, Lima: Sinco Editores, 342–362.

Vivas, Reyna, Eugenio Tarazona, Carlos Caballero, and Nancy Marrero (1988). *El Sistema Nacional de Salud: Administración, Presupuestación, Gasto, y Financiamiento*, Bogotá: Fedesarrolllo.

Waterman, Richard W., and Kenneth Meier (1998). "Principal-Agent Models: An Expansion?" *Journal of Public Administration Research and Theory*, 8:2, 173–202.

Webb, Richard (1994). "Peru," in John Williamson (Ed.), *The Political Economy of Policy Reform*, Washington: Institute for International Economics.

Webb, Richard, and Sofía Valencia (2006). "Human Resources in Public Health and Education," in Daniel Cotlear (Ed.), *New Social Contract for Peru*, Washington: World Bank.

Weber, Max (1946). "Bureaucracy," in Max Weber, *Essays in Sociology*, Hans Heinrich Gerth (Ed.) and C. Wright Mills, (Tran.), New York, Oxford University Press.

Weyland, Kurt (2006). *Bounded Rationality and Policy Diffusion: Social Sector Reform in Latin America*, Princeton: Princeton University Press.

(2005). "Theories of Policy Diffusion: Lessons from Latin American Pension Reform," *World Politics* 57:2, 265–295.

(2002a). *The Politics of Market Reform in Fragile Democracies: Argentina, Brazil, Peru, and Venezuela*, Princeton: Princeton University Press.

(2002b). "Limitations of Rational Choice Institutionalism for the Study of Latin American Politics," *Studies in Comparative International Development*, 37:1, 57–85.

(1998). "The Political Fate of Market Reform in Latin America, Africa, and Eastern Europe," *International Studies Quarterly*, 42:4, 645–673.

(1996). *Democracy without Equity*, Pittsburgh: University of Pittsburgh Press.

Wilde, Alexander (1978). "Conversations among Gentlemen: Oligarchical Democracy in Colombia," in Juan J. Linz and Alfred Stepan (Eds.) *The Breakdown of Democratic Regimes: Latin America*, Baltimore: Johns Hopkins University Press.

Williams, Mark Eric (2006). "Escaping the Zero-Sum Scenario: Democracy versus Technocracy in Latin America," *Political Science Quarterly*, 121:1, 119–139.

Wise, Carol (2003). *Reinventing the State: Economic Strategy and Institutional Change in Peru*, Ann Arbor: University of Michigan Press.

(1986). "Economía Política del Perú: Rechazo a la Receta Ortodoxa," Documento de Trabajo N. 16. Lima: Instituto de Estudios Peruanos.

World Bank (2004). *World Development Report 2004: Voice of Users, or the Poor*, Washington, DC: World Bank.

(2000). *Bolivia: From Patronage to a Professional State*, Report No. 20115-BO, Washington, DC: World Bank.

(1993). *World Development Report 1993: Investing in Health*. Washington, DC: World Bank, New York: Oxford University Press.

(1987). *Financing Health Services in Developing Countries: An Agenda for Reform*, Washington, DC: World Bank.

Yepes, Francisco, Manuel Ramírez, Luz Helena Sánchez, Marta Lucía Ramírez, and Iván Jaramillo (2010). *Luces y Sombras de la Reforma de Salud en Colombia: Ley 100 de 1993*. Bogotá: Universidad del Rosario/Assalud/Mayol Ediciones.

Yglesias, Arturo (1999). "Marco de Análisis para la Evaluación y el Diseño e Implementación de las Reformas de Atención en Salud," in Felipe Portocarrero (Ed.), *Pobreza y Políticas Sociales en el Perú: Nuevos Aportes*, Lima: Red para el Desarrollo de las Ciencias Sociales.

Interviews

Colombia

Arenas, Roberto (2006). Author interview. Director of the NPD (1970–1974). Bogotá: 10th May.

Arias, Andrés (2009). Author interview. Minister of Agriculture (2006–2010) and Vice Minister of Finance (2003–2004). Candidate, conservative party presidential primaries (2010). Bogotá: 28th July.

Ávila, Ricardo (2008). Author interview. Economic adviser to President César Gaviria (1990–1994). Editor of *Portafolio*. Bogotá: 7th May.

Barco, Víctor Renán (2006). Author interview. Senator. Member of the Budget Commission. Bogotá: 17th May.

Botero, Rodrigo (2009). Author telephone interview. Minister of Finance (1974–1976). Founder of Fedesarrollo. Boston: 27th July.

Caballero Argáez, Carlos (2007). Author interview. Member, Monetary Board (2001–2003). Dean, Public Policy Department, Los Andes University. Bogotá: 8th October.

Cano, Carlos Gustavo (2008). Author interview. Director, Central Bank. Bogotá: 29th April.

Cadena, María Isabel (2009). Author interview. General Secretary, Ministry of Finance. Bogotá: 23rd July.

Cárdenas, Jorge Hernán (2008). Author interview. Expert in administrative reform. Bogotá: 25th January.

Cárdenas, Mauricio (2008). Author interview. NPD Director (1999–2000). Director, Fedesarrollo. Bogotá: 21st January.

Castaño, Ramón (2008). Author interview. Health expert. Member and Director of PARS. Bogotá: 8th February.

Cepeda, Manuel (2006). Author interview. Constitutional Justice. President, Constitutional Court (2001–2009) Bogotá: 18th May.

Chávez, Juliana (2008). Author interview. Adviser, office of the Vice President. Assistant in the DNP (2001–2005). Bogotá: 8th February.

Dávila, Andrés (2008). Author interview. Political Scientist and Adviser, office of the Vice President. Chief of the Justice and Security area, DNP, 2001–2007, Bogotá: 8th February.

Echeverry, Juan Carlos (2008). Author interview. NPD Director (2000–2002). Minister of Finance (2010–). Bogotá: 24th January.

Galán Sarmiento, Augusto (2009). Author interview. Minister of Health (2005–2006). Bogotá: 29th July.

Gaviria, Alejandro (2008). Author interview. Deputy Director of NPD (2002–2004). Dean, Economics Department, Universidad de Los Andes. Bogota: 4th March.

Gaviria, César (2013). Author telephone interview. Colombian President (1990–1994). Bogotá: 7th October.

Giedion, Úrsula (2008). Author interview. Health expert. Consultant to the Ministry of Health and PARS. Bogotá: 11th February.

Hernández, Gregorio (2008). Author interview. General Manager, Banco de la República. Bogotá: 7th May.

Hommes, Rudolph (2008). Author interview. Minister of Finance (1990–1994). Bogotá: 28th May.

Janna, Michel (2013). Director of Public Credit, Ministry of Finance (April 2013–). Bogotá: 24th September.

Junguito, Roberto (2008). Author interview. Minister of Finance (1984–1986; 2002–2003). Monetary Bank Board (1991–1999). Bogotá: 7th February.

Kalmanovitz, Salomón (2008). Author interview. Member, Monetary Board of Banco de la República (1993–2005). Bogotá: 24th April.

La Serna, Juan Mario (2008). Author interview. Member, Monetary Board of Banco de la República, (2005–2009). Bogotá: 29th April.

Llorente, Rodrigo (2006). Author interview. Minister of Finance (1973–1974). Bogotá: 15th May.

Mejía, Luis Bernardo (2006). Author interview. NPD official. (1996–2001). Bogotá: 14th May 2006.

Montenegro, Santiago (2008). Author interview. NPD Director (2002–2006). Bogotá: 28th February.

Morales Benites, Olympo (2008). Author interview. Politician. Member, liberal party and *Alas Equipo Colombia*. Bogotá: 2nd May.

Morales Benites, Otto (2006). Author interview. Liberal party leader. Minister of Agriculture, (1960–1962, Minister of Labor (1958–1960). Bogotá: 15th May.

Ordóñez, Antonio (2006). Author interview. Minister of Health (1966–1970). Health expert. Bogotá: 17th May.

Paredes, Nelcy (2008). Author interview. Health expert. Member, "Team Change." Adviser, ACEMI. Bogotá: 20th February.

Perry, Guillermo (2008). Author interview. Minister of Finance (1994–1996). Bogotá: 26th February.

Pinto, Diana (2009). Author interview. Health expert. Member, PARS. Adviser, Ministry of Social Protection. Researcher, Fedesarrollo. Bogotá: 23rd July.

Ramírez, Jaime (2009). Author interview. Health expert. Senior technical Advisor of PARS. (1999–2001). Bogotá: 28th July.

Reina, Mauricio (2008). Author interview. Vice Minister of Commerce (1994–1995). Researcher, Fedesarrollo. Bogotá: 27th February.

Reyes Forero, Carlos (2012). Author phone interview. Colombian health expert. Adviser to Minister Mauricio Santa María. Bogotá: 17th January.
Samper, Ernesto (2008). President of Colombia (1994–1998), Bogota 14th February.
Santa María, Mauricio (2008). Author interview. Deputy Director, NPD (2005–2007). Minister of Social Protection (2010–2012). Bogotá: 28th February.
Steiner, Roberto (2009). Author interview. Director of Fedesarrollo (2009–). Bogotá: 21st July.
Tono, Teresa (2008). Author interview. Health expert. Member and Director of PARS (1996–2008). Adviser, Ministry of Health. Bogotá: 24th April.
Uribe, Juan Pablo (2008). Author interview. Health expert. Adviser, Ministry of Health. Vice Minister of Health (2002). Bogotá: 29th January.
Urrutia, Miguel (2008). Author interview. Director, NPD (1974–1977). Adviser to the Monetary Board (1966–1970). Secretary General, Banco de la República (1993–2005). Bogotá: 25th February.
Vidal Perdomo, Jaime (2006). Author interview. Secretary of the Presidency (1958–1962). Bogotá: 13th May.
Zapata, Juan (2011). Author phone interview. Colombian health expert. Member, Fedesarrollo. Bogotá: 16th November.

Peru

Abugattás, Javier (2007) Author interview. Vice Minister of Economy (1988). Adviser, MEF (various periods over the last twenty years). Lima: 10th August.
Abusada, Roberto (2010) Author interview. Viceminister of Commerce in MEF (1980–1982) and advisor in MEF during some periods during the nineties. Lima 29th March.
Alva Orlandini, Javier (2007) Author interview. Vice President (1980–1985). Senator (1980–1990). Leader and member of *Acción Popular*. Lima: 20th June.
Álvarez, Rosalía (2007). Author interview. Secretary General, MEF (2006–2008). Official in MEF since 1982. Lima: 4th December.
Álvarez Rodrich, Augusto (2007) Author interview. Member, Board of OSIPTEL (1993–1998). Lima: 15th June.
Aráoz, Mercedes (2013). Author telephone interview. Minister of Economics (December 2009–September 2010) Lima: 29th November.
Arce, Melitón (2007). Author interview. Vice Minister of Health (1985–1988; 2007). Adviser, MINSA (2006–2007). Lima: 18th June.
Arias Stella, Javier (2007). Author interview. Minister of Health (1964–1966), (1967–1968). Minister of Foreign Affairs (1980–1983). Lima: 17th July.
Arrieta, Alejandro (2007) Author interview. Adviser, MEF (various periods since 1980). Lima: 3rd June.
Arrisueño, Gabriel (2007) Author interview. Researcher, Institute for Liberty and Democracy (ILD). Lima: 15th August.
Arrunátegui, Jorge (2013) Author interview. Adviser, MEF (2003–2006). Vice Minister of Social Benefits, Ministry of Development and Social Inclusion (2011–2013). Lima: 30th December.
Ballén, Mariana (2007). Author interview. Adviser, MEF. Lima: 5th December.
Barletti, Bruno (2007). Author interview. Adviser, MEF. Official in charge of implementing the Sistema Integrado de Administración Financiera (SIAF). Lima: 20th July.

Blume, Cecilia (2007). Author interview. Chief Adviser, MEF (2005–2006). Lima: 7th July.
Boza, Beatriz (2007). Author interview. President of INDECOPI (1994–2000). Adviser, MEF (2000–2001). Lima: 30th July.
Burneo, Kurt (2007). Author interview. Vice Minister of Finance (2001–2004), MEF. Member, Board of the Central Bank (2001–2006). Lima: 15th June.
Cáceres, Rosario (2007). Author interview. Chief, Parliamentary Coordination Office, MEF. Lima: 2nd August.
Calderón, José (2007). Author interview. Vice Minister of Health (2007). Doctor in MINSA (since the 1980s). Lima: 21st June.
Campos, Fernando (2007). Author interview. Legal Adviser, MINSA. Negotiator with unions. Lima: 21st June.
Carranza, Luis (2010). Author interview. Minister of Economics (2006–2008; 2009). Lima: 11th February.
Carrillo, Carlos (2007). Author interview. Director (1990–1992), (1994–1999), Instituto Nacional de Salud. Lima: 16th October.
Casas, Carlos (2013). Author interview. Vice Minister of Economics (2010). Chief of Advisers (2006–2008), and Adviser (1999), MEF. Lima: 21st November.
Chávez, Benjamín (2008). Author interview. Adviser, Ministry of Commerce. Lima: 16th November.
Cortez, Juan Carlos (2007). Author interview. Deputy Director, Ciudadanos al Día. Lima: 2nd July.
Cubas, Víctor (2007). Author interview. Vice Minister of Health (1990–1991). Dean, Public Health Department, Universidad Cayetano Heredia. Lima: 3rd July.
Del Castillo, Jorge (2007). Author interview. President, Council of Ministers (2006–2008). Lima: 28th June.
Díaz, Álvaro (2007). Author interview. Adviser and Negotiator (Free Trade Agreements), Ministry of Commerce. Lima: 16th November.
Du Bois, Fritz (2007). Author interview. Chief of Cabinet, MEF (1991–1997). Lima: 18th June.
Escaffi, José (2007). Author interview. General Manager of OSIPTEL (1993–1997). Lima: 15th July.
Espinoza, Rubén (2007). Author interview. Consultant and Researcher, Instituto Nacional de Salud. Adviser, MINSA (1990s). Lima: 16th August.
Farah, Eduardo (2007). Author interview. President, National Society of Industry (2007). Lima: 27th June.
Ferrero, Carlos (2007). Author interview. President, Council of Ministers (2003–2005). Lima: 1st June.
Ferraro, Carlos (2007). Author interview. Vice Minister of Commerce (2002–2009). Lima: 30th October.
Francke, Pedro (2007). Author interview. Director, FONCODES (2001–2003). Expert in health policies. Lima: 19th June.
Fuentes Rivera, José (2007). Author interview. MINSA Official in Directive Positions since the eighties. Lima: 4th July.
Gálvez, Nelly (2007). Author interview. Official, DIGEMID (Medical drugs regulation office in MINSA). Lima: 17th August.
Gana Perú (2011). La Gran Transformación: Plan de Gobierno 2011–2016, Lima: Gana Perú.

García, Uriel (2007). Author interview. Minister of Health (1980–1982). Lima: 10th July.

Gastañeta, Rafael (2007). Author interview. Adviser, Minister of Commerce. Lima: 16th November.

Gozzer, Ernesto (2007). Author interview. Consultant, MINSA (1991–1993). Member, Office of International Cooperation (1994–1999). Lima: 21st August.

Guerra García, Roger (2007). Author interview. Congressman 1995–2000. Dean, Universidad Cayetano Heredia (1989–1994). Lima: 9th July.

Guerrero, Luis (2007). Author interview. Official, SIAF, MEF. Lima: 11th December.

Hume, María Jesús. (2007). Author interview. Vice Minister of Commerce (1980–1982), MEF. Lima: 13th December.

Iwsaki, Maria Lila (2007). Author interview. Advisor and Secretary General (1999) in MEF. Lima, 27 September.

Johnson, Jaime (2007). Author interview. Public Health Expert. Adviser, MINSA and other social policy agencies. Lima 26th June.

Kapsoli, Javier (2007). Author interview. Director, Social and Economic Affairs, MEF (2007). Lima: 11th June.

Kuczynski, Pedro Pablo (2007). Author interview. Minister of Economics (2001–2002; 2004–2005). President, Council of Ministers (2005–2006). Lima: 17th August.

León García, Luis (2007). Author interview. Director, General Health Office, MINSA (2007). Lima: 19th October.

Lescano, Yonhy (2007). Author interview. Congressman for Puno. Lima: 5th December.

Llanos, Alejandro (2007). Author interview. Public health expert. Director, Epidemiology Office, National Health Institute (1985–1992). Adviser in the Andean Pact. Lima: 3rd July.

Lombardi, Guido (2008). Author interview. Congressman, Unidad Nacional (2006–2011). President, Economy Committee, Peruvian Congress. Lima: 17th August.

Luján, Manuel (2007). Author interview. Doctor in MINSA (1982–present). Vice Ministerial Adviser (2004–2006). Lima: 26th July.

Marchena, Pedro (2007). Author interview. Expert in MINSA (1990s). Lima: 24th October.

Martinez, Juan José (2007). Author interview. Legal Adviser, MEF. Lima: 21st November.

Mazetti, Pilar (2007). Author interview. Minister of Health (2004–2006). Lima: 29th November.

Meloni, Augusto (2007). Author interview. Public health expert. Director, Office of Cooperation (1993–1999). Lima: June 1st and 4th August.

Mendoza, Pedro (2007). Author interview. Member, International Cooperation Office, MINSA (1995–2000). Lima: 22nd August.

Mendoza, Waldo (2007). Author interview. Vice Minister of Finance (2004–2005). Lima: 6th June.

Monteagudo, Manuel (2008). Author interview. Legal Adviser, Central Bank. Lima: 11th August.

Montoya, Andrés (2007). Author interview. Adviser, MEF (2003–2004). Lima: 6th July.

Montoya, Víctor Hugo (2007). Author interview. Legal Expert, Constitutional Tribunal. Lima: 6th December.

Morales, José Miguel (2007). Author interview. President, CONFIEP (2005–2006). Lima: 25th June.
Mori, Antonio (2007). Author interview. MEF official. Member, SIAF, MEF. Lima: 11th December.
Morón, Eduardo (2007 and 2013). Author interview. Vice Minister of Economics (2008–2009), Adviser, MEF and MINCETUR. Researcher, Center of Research, Universidad del Pacífico (CIUP). Lima: 14th November 2007; 19th November 2013.
Nakasone, Eduardo (2007). Author interview. Adviser, MEF (2003–2004). Lima: 17th July.
Negrón, Carmen (2007). Author interview. MEF official. Lima: 30th May.
Palacios, Rosa María (2007). Author interview. Adviser, PCM. Journalist. Lima: 10th July.
Pasco Font, Alberto (2007). Author interview. Economist. Executive Director, COPRI. Lima: 3rd December.
Pastor, Ada (2007). Author interview. Deputy Director, PRAES. Public Officer, Instituto Peruano de Seguridad Social (IPSS) now ESSALUD. Lima: 19th October.
Peña, Manuel (2007). Author interview. Director, Pan American Health Organization (PAHO), Peru. Lima: 6th August.
Pérez, Martín (2008). Author interview. Minister of Commerce (2008–2010). Lima: 19th August.
Pinillos, Luis (2007). Author interview. Minister of Health (1988–1989). Lima: 21st June.
Polarollo, Pierina (2007). Author interview. Adviser, MEF. Health reform expert. Lima: 14th June.
Portocarrero, Augusto (2010). Author interview. Health expert. Ministry Adviser and Health Official, MINSA. Lima: 5th August.
Pretell, Eduardo (2007). Author interview. Minister of Health (2000–2007). Lima: 27th November.
Prialé, Miguel (2007). Author interview. Director, SNIP. Lima: 6th June.
Quijandría, Jaime (2009). Author interview. Minister of Economics (2003–2004). Lima: 25th June.
Ramírez, Gladys (2007). Author interview. Deputy Director, Epidemiology Office, Ministry of Health. Lima: 28th May.
Rettberg, Angelika (2009). Author interview. Political scientist. Expert on business groups in Colombia. Bogotá: 24th July 2009.
Rivera, Iván (2011). Minister of Industry (1983–1984). Advisor in MEF (1993–1996). Skype Interview. Washington 4th January.
Salazar, Ricardo (2007). Author interview. President, CONSUCODE (2001–2003). Lima: 13th June.
Sánchez Moreno, Francisco (2007). Author interview. Vice Minister of Health (1978–1979). Director, Academia Peruana de la Salud. Lima: 13th August.
Seinfield, Janice (2007). Author interview. Public health expert. Researcher, Centro de Investigación de la Universidad del Pacífico. Lima: 14th November.
Solari, Luis (2007). Author interview. Health Minister (2001–2002). First Minister (2002–2003). Congressman, Peru Posible (2001–2006). Lima: 11th June.
Sotelo, Eduardo (2007). Author interview. Legal adviser, MEF and SUNAT. Lima: 21st August.

Távara, José (2007). Author interview. Director, OSIPTEL (2004–2007). Vice Minister of Communications (2004–2007). Lima: 6th December.

Tejada de Rivero, David (2007). Author interview. Minister of Health (1985–1987; 1989). Lima: 10th July.

Teullet, Patricia (2007). Author interview. Vice Minister of Economics (2001–2002). Lima: 12th June.

Ugarte, José Miguel (2008). Author interview. Director, National Direction of Public Indebtment, MEF. Lima: 7th August.

Ugarte, Mayén (2006). Author interview. Adviser, MEF (1994–1998). Lima: 10th June.

Ugarte, Óscar (2007). Author interview. Minister of Health (2009–2011) and Vice Minister of Health (2001–2002). Lima: 10th July.

Valencia, Sofía (2007). Author interview. Economist. World Bank Consultant. Lima: 5th July.

Vásquez, María Inés (2007). Author interview. Adviser, MEF (2003–2006). Lima: 25th May.

Vera, José Carlos (2007). Author interview. Adviser, MEF and MINSA (1993–1994). Lima: 17th October.

Vidal, Álvaro (2007). Author interview. Minister of Health (2003–2004). Lima: 14th August.

Vílchez, Nidia (2007). Author interview. Congresswoman for Junín. Member, Budget Commission, Congress. Lima: 6th December.

Villarán, Fernando (2007). Author interview. Minister of Labor (2001–2003). Lima: 26th July.

Von Hesse, Milton (2007). Author interview. Adviser, MEF. Director of SNIP (2002–2006). Lima: 14th June.

Webb, Richard (2007). Author interview. President, Central Bank (1983–1985; 2001–2003). Lima: 6th December.

Zamalloa, Edgar (2007). Author interview. Adviser, MEF (1998–2004). Director of Economic Analysis, MEF (2005–2006). Lima: 17th July.

Zavala, Fernando (2007). Author interview. Minister of Economics (2005–2006). Lima: 15th October.

Zegarra, Gonzalo (2007). Author interview. Editor, *Semana Económica*. Lima: 31st May.

Index

An *n* following a page number indicates a note; a *t* following a page number indicates a table.

For EU product safety concerns, contact us at Calle de José Abascal, 56–1°,
28003 Madrid, Spain or eugpsr@cambridge.org.

www.ingramcontent.com/pod-product-compliance
Lightning Source LLC
LaVergne TN
LVHW011521200926
848484LV00001B/100

* 9 7 8 1 1 0 7 0 5 9 8 7 0 *